D0191296

PENGUIN BOOKS

Who Ate the First Oyster?

Cody Cassidy is the coauthor of the popular science book *And Then You're Dead*, which was translated into more than ten languages, and a former bookstore clerk in Buenos Aires. He lives in San Francisco.

Who Ate the First Oyster?

The Extraordinary People Behind
the Greatest Firsts in History

CODY CASSIDY

PENGUIN BOOKS

PENGUIN BOOKS
An imprint of Penguin Random House LLC
penguinrandomhouse.com

LIBRARY OF CONGRESS CATALOGING-IN-PUBLICATION DATA
Names: Cassidy, Cody, author.
Title: Who ate the first oyster? : the extraordinary people behind the
greatest firsts in history / Cody Cassidy.
Description: [New York] : Penguin Books, [2020] |
Includes bibliographical references.
Identifiers: LCCN 2019030541 (print) | LCCN 2019030542 (ebook) |
ISBN 9780143132752 (paperback) | ISBN 9780525504672 (ebook)
Subjects: LCSH: History—Miscellanea. | Biography—Miscellanea.
Classification: LCC D10 .C336 2020 (print) |
LCC D10 (ebook) | DDC 904/.7—dc23
LC record available at https://lccn.loc.gov/2019030541
LC ebook record available at https://lccn.loc.gov/2019030542

Printed in the United States of America
5 7 9 10 8 6 4

Set in Baskerville MT Std
Designed by Cassandra Garruzzo

To Mom and Dad

Three-Million-Year Timeline

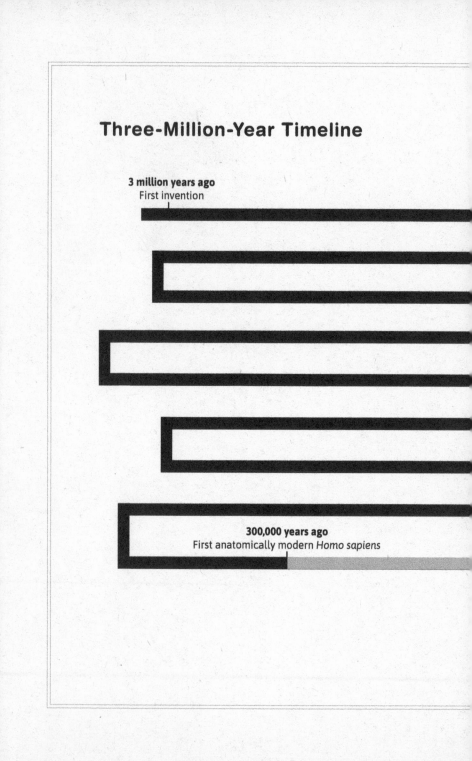

3 million years ago
First invention

300,000 years ago
First anatomically modern *Homo sapiens*

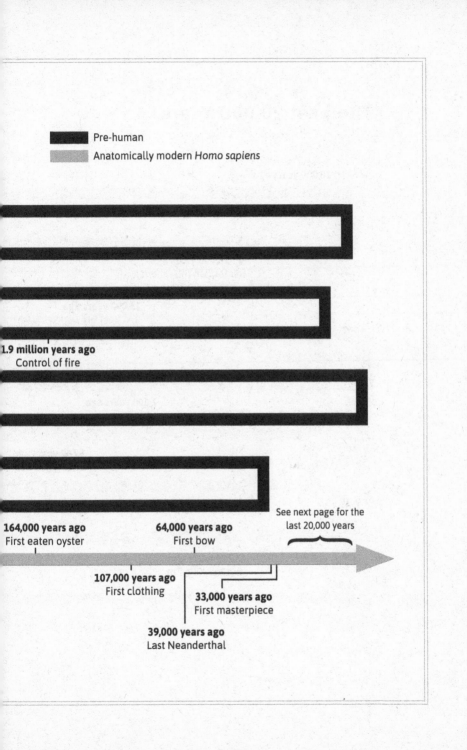

Pre-human

Anatomically modern *Homo sapiens*

1.9 million years ago
Control of fire

164,000 years ago
First eaten oyster

64,000 years ago
First bow

See next page for the
last 20,000 years

107,000 years ago
First clothing

33,000 years ago
First masterpiece

39,000 years ago
Last Neanderthal

The Last 20,000 Years

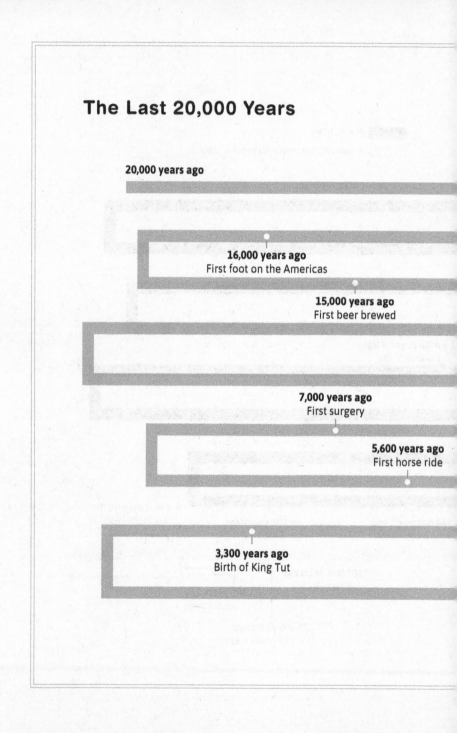

20,000 years ago

16,000 years ago
First foot on the Americas

15,000 years ago
First beer brewed

7,000 years ago
First surgery

5,600 years ago
First horse ride

3,300 years ago
Birth of King Tut

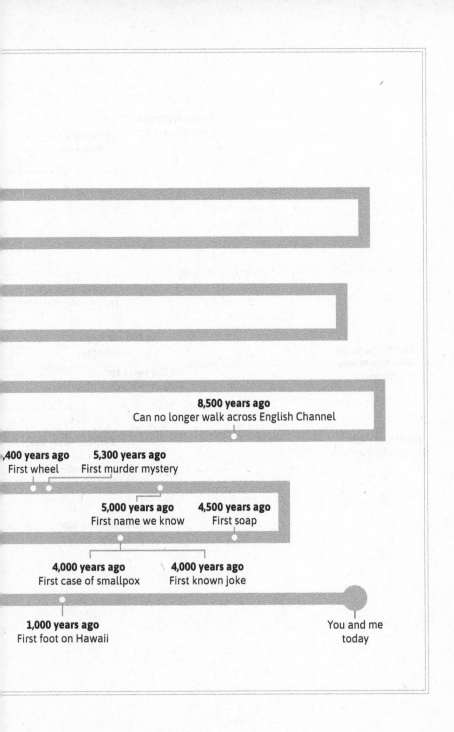

8,500 years ago
Can no longer walk across English Channel

400 years ago
First wheel

5,300 years ago
First murder mystery

5,000 years ago
First name we know

4,500 years ago
First soap

4,000 years ago
First case of smallpox

4,000 years ago
First known joke

1,000 years ago
First foot on Hawaii

You and me
today

33,000 years ago
Chauvet cave art

7,000 years ago
First surgery _____

5,300 years ago
First murder mystery

39,000 years ago
Last Neanderthal _____

15,000 years ago _____
First beer brewed

16,000 years ago
First foot on the
Americas

1,000 years ago
First foot on Hawaii

3 million years ago
The invention of
inventions

164,000 years ago _____
First eaten oyster

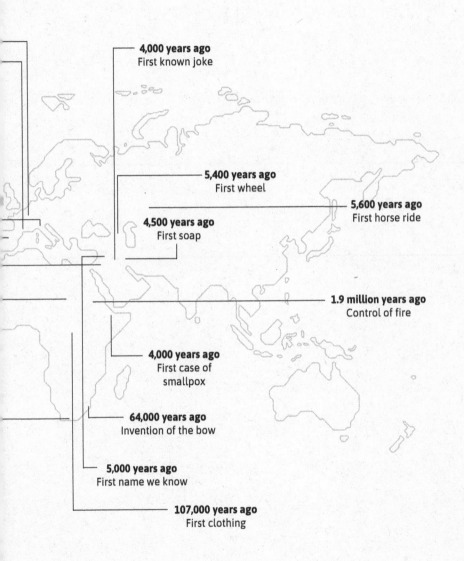

4,000 years ago
First known joke

5,400 years ago
First wheel

5,600 years ago
First horse ride

4,500 years ago
First soap

1.9 million years ago
Control of fire

4,000 years ago
First case of
smallpox

64,000 years ago
Invention of the bow

5,000 years ago
First name we know

107,000 years ago
First clothing

Contents

Introduction

He was a bold man that first ate an oyster.

JONATHAN SWIFT

In 1991, the victim in the world's most interesting murder case was found 10,500 feet above sea level in the Ötztal Alps in northeastern Italy, 15 feet from the Austrian border. Dubbed Ötzi, the man had been shot with an arrow in the back nearly 5,300 years ago, and his body has since become the most carefully studied corpse in human history. In the fall of 2017 I decided to visit the murder scene. Though this was my first criminal investigation, I began as I presumed any good homicide detective would: I retraced the victim's last steps.

Remarkably, even though the murder occurred nearly one thousand years before the construction of the Great Pyramid, this retracing is actually possible. Thanks to scientists identifying layers of pollen from the victim's digestive system as well as their sources, we now have an accounting of Ötzi's final twelve hours far more accurate than any bloodhound could provide.

Ötzi's last hike took place in what is now a piece of northern

Italy, sliced off from Austria after World War I, though when I visited, it seemed unclear whether anyone had ever told the people who live there. The architecture, the food, the culture, the signs, and even the greetings were so comprehensively Austrian I checked a map to make sure I hadn't crossed the border.

I began my trek early in the morning, and it soon became clear that Ötzi must have been in fine shape on the day he died. The Ötztal Alps do not rise slowly like the foothills of the Sierra Nevada Mountains I was used to. Instead, they rocket out of river valleys at such steep angles even the gentler path Ötzi chose was crisscrossed in sharply angled switchbacks that rose into the snow and fog.

Investigators have established that Ötzi died shortly after enjoying a leisurely lunch at the top, which suggests he was a far better meteorologist than I. Snow had begun to fall and dense fog blanketed the pass when I arrived at the peak, and as I contemplated the tricky traverse to his final resting place, I spotted a few mountaineers—the first I had seen all day—strapping into crampons. We didn't share many words in common, but after a few gestures toward my tennis shoes we did share an understanding that if I continued I was at some risk of making Ötzi's final resting place my own. Less than a quarter mile from the site of the murder, and six thousand miles from home, I decided, in this case, that interviews with archaeologists who had investigated the scene would have to suffice.

The aborted trip to the murder site was a part of an expansive, three-year-long project to produce this book. It began as an inquest into humankind's greatest "firsts" but quickly expanded to include profiles of the individuals responsible. The more I learned about prehistoric discoveries, the more I wanted to know the people who made them. Yet most reconstructions of the prehistoric ignore the existence of individuality entirely, and speak of "peoples" rather than people.

So I set out to find remarkable *people* from our deep history. I interviewed more than one hundred experts, read dozens of books and hundreds of research papers. I ordered obsidian off the internet and tried to shave my face with it. I visited the site of humankind's first great piece of art. I started a fire with flint and pyrite. I fired a replica of an ancient bow. I spoiled gruel to brew beer. And I quite nearly joined Ötzi in his final resting place.

In the end, I identified seventeen ancient individuals who lived before or without writing. These are people who scholars know existed and whose extraordinary or fateful acts are the foundation of modern life. Then I asked everyone from archaeologists to engineers, geneticists to lawyers, and astrologists to brewmasters who these anonymous individuals might have been, what they were thinking, where they were born, what they spoke (if they spoke!), what they wore, what they believed, where they lived, how they died, how they made their discovery, and most important, why it mattered.

When viewed from the distance of many thousands of years, cultural, technological, and evolutionary change appears to proceed in a smooth line. Stone tools gradually give way to metal; furs gradually give way to woven fabric; gathered berries gradually give way to cultivated crops. Because of the appearance of a slow gradation, it's tempting to assume that no single individual could possibly have played a significant role in the seemingly inevitable trajectory of human history—or the seemingly glacial pace of human evolution.

But this gradation is the illusion of our perspective. It neglects the way technology and even evolution have always occurred: in fits and starts, with individuals at the forefront. Rolling logs do not inevitably transition into wagons. Instead, someone invented the wheel and axle—regarded by many scholars as the greatest mechanical invention of all time—and someone fired the first bow and arrow—probably the most successful weapon system the world has ever seen. Thanks to the imperfect reach of written history, we've lost their names, but a name is a detail, and modern science now provides far more revealing details about the geniuses of the prehistoric.

———

Those two words—"genius" and "prehistoric"—are not often put together thanks to the stereotypes of cartoons, early caricatures, and the temptation to mistakenly equate tools and technology with intelligence. Though "prehistoric" is supposed to refer only to those who lived before writing, its first listed synonym is "primitive" and the implications are clear: The people who lived "before the dawn of history" were illiterate savages. Morons. Brutes who lived in dark caves, munching on mammoth burgers between grunts.

But like most stereotypes, this one collapses under even the briefest interrogation. The so-called cavemen—who for the most part didn't even live in caves—required a far wider knowledge base than those of us living in the era of mass food production and job specialization. Their survival depended upon an encyclopedic understanding of their environment. They each had to find, gather, hunt, kill, and craft virtually everything they ate, lived in, or used. They had to know which plants killed you, which ones saved you, which ones grew in what seasons and where. They had to know the seasonal migration patterns of their prey. According to the scholars I spoke with there's no evidence geniuses were any less common in ancient history than today, and at least some evidence that they were more so.

It feels controversial, or even speculative, to assert that geniuses lived in prehistoric times. It shouldn't be.

Just as prehistoric people had their fair share of nitwits, buffoons, dopes, traitors, cowards, scallywags, and evil, revenge-seeking psychopaths—a few of whom I discuss in the following pages—so too were there the equivalents of da Vincis and Newtons. That isn't just speculation. It's a provable, verifiable, indisputable fact. The evidence is brushed on cave walls in France, scratched into clay tablets in the Middle East, found on islands in the South Pacific, and buried on top of four wheels in Russia. If Newton is feted for inventing calculus, what should we think of the person who invented math itself? If Columbus is celebrated for stumbling upon the Americas, what should we think about the person who actually did discover them sixteen thousand years earlier? And what of the person who searched for and found the world's most isolated archipelago five hundred years before Columbus (accidentally) found a continent?

"Prehistoric" simply means that their names and stories went unrecorded and nothing more. Their lives were no less remarkable than those who lived afterward and, in at least a few cases, far more so.

Common sense should have dictated this long ago. Modern science has removed all doubt.

Until now, little has been written about these ancient individuals partly because there was so little to say. Early archaeologists found bones and tools, but not enough to speak to the humanity, individuality, and motives of their owners.

But within the past few decades, modern science has illuminated our ancient past to a startling degree. Thanks to techniques for recovering and analyzing DNA, ancient bones tell astonishing new stories—stories about the survivors who lived at the edge of the habitable world, the origins of plagues, and even the invention of clothing. Paleolinguists have reconstructed ancient languages to trace population movements, lifestyles, even the location of some inventions—including perhaps, the home of the wheel itself.

Old fashioned archaeology has also undergone a dramatic change in the last two decades. The number of discoveries has exploded to the degree that authors invariably include a plea for forgiveness for the inevitable revelations that will occur in the waiting period between writing and publishing (consider this mine). Writing about prehistory has become a game of whack-a-mole, and not just because of new finds, but because of the new tools applied to old ones.

Recent anthropological studies have even revealed the mind-sets of these ancient people. Studies by scholars like the University of Santa Barbara's Donald Brown have exposed remarkable consistencies across hundreds of human cultures as seemingly different as the highlanders of Papua New Guinea and the bankers on the streets of lower Manhattan. Brown and others' search for similarities have yielded a list of what anthropologists call "human universals," a revealing and peculiarly specific set of traits exibited by *every* culture.

When Marco Polo returned from his thirteenth-century voyage, he shocked Europe with his tales of the neck elongation practiced by the Padaung and Kayan peoples of Thailand and Myanmar. But while neck elongation and Western bow ties might seem to be the product of two vastly different mind-sets, they stem from the universal human desire for individualization and body decoration. It would have been far stranger if Polo had discovered a culture in which no one decorated themselves—yet no anthropologist has done so. Body decoration is one of hundreds of human universals that anthropologists like Brown have identified, and many researchers believe these universals offer the best lens through which to view ancient cultures whose archaeological remains haven't survived. They do not describe individuals, but they help describe what it is to be human.

Despite the powerful tools we now use to examine our deep past, many fundamental questions remain. When I asked two of the world's leading archaeologists when *Homo sapiens* began to speak full languages and think like modern humans, their answers differed by more than one hundred thousand years. Such is the stubborn opaqueness of our past.

Nevertheless, with modern tools, scholars can now engage in more educated speculation about the greatest people, moments, and firsts of ancient human history than ever before.

I had previously pondered humankind's peculiar firsts—as I suspect many of us have when trying something new and particularly bizarre—but I didn't consider the questions deeply until I read of a poignant note written by an ancient Egyptian physician describing a tumor of the breast in a patient of his. Historians believe it's the first documented case of cancer. At the end of a long and detailed description of the spreading tumor, the physician simply adds: "There is no treatment."

I found something touching in the specificity of this ancient woman suffering from this ancient disease. A specificity, and an individuality, that I found lacking in the typical descriptions of ancient "peoples." So I set out to find out about not just humankind's ancient firsts, but about the people who accomplished them.

This is a book about who these people were. What they did. And why it mattered.

A quick note on the large numbers problem

You, me, and everyone up to and including academics cannot truly fathom great lengths of time. We perceive the difference between twenty thousand and thirty thousand or three hundred thousand years simply as "a long time ago." The information is far too impractical, abstract, and ultimately useless for us to comprehend. It's the same problem astronomers run into when discussing the vastness of space, or the weight of a white dwarf. Our minds are not built to deal with this level of abstraction, and the tendency when confronted with one of these unimaginably large numbers is to use words like "really big," or "really far," or "really long ago," and quickly move on.

The problem, of course, is that even though these large numbers of years seem the same to us, they're not. And when it comes to human history, the differences have enormous meaning. Recent archaeological finds, for example, have revealed *H. sapiens* overlapped territorially with Neanderthals for more than five thousand years. That's the same length of time as *all of written history*. Properly evaluating the difference between one thousand and five thousand years should deeply complicate the commonly held belief in the smart *H. sapiens* steamrolling the Neanderthals out of existence, which in turn should force us to reevaluate the short shrift commonly given to our ancestral cousin's intellectual abilities.

If we're to understand the relationships between different species, cultures, and the nature of our evolution, we have to feel the differences in these massive numbers. So I have contrived a few ways to help. For a period of time, I thought about putting dollar signs in front of the years, so what was once 1,000 became $1,000, in an attempt to make the abstract into something tangible. Instead, I have opted for the familiar countdown of a clock (with the notable exception of the first two chapters, which took place long before humans evolved). If a single day represented the three hundred thousand years since the evolution of anatomically modern *H. sapiens*, written history would begin a half hour before midnight. That leaves twenty-three and a half hours of "prehistory," a place that an estimated 1.5 billion anonymous people called home.

In light of the big numbers problem, I have avoided citing large numbers of years except when absolutely necessary, preferring in their stead to provide the context in which the people lived. But sometimes discussing unintelligibly large numbers is unavoidable. In these instances, it may help you—as it does me—to think dollars, timelines, or even hours, instead of years.

Who
Ate the
First
Oyster?

Who Invented Inventions?

This occurred 3 million years ago,
which is before humans evolved.

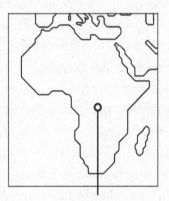

3 million years ago
The invention of inventions

In October 1960, a then twenty-six-year-old Jane Goodall observed a chimpanzee she dubbed David Greybeard strip a long twig of its leaves, use it to probe a termite mound, and lick away the bugs he retrieved. It may have been just a snack for Greybeard, but to a scientific community who at the time defined *Homo sapiens* by their unique use of tools, it was earth shattering. Goodall immediately telegraphed the news to the paleoanthropologist

Louis Leakey, who famously responded, "Now we must redefine tool, redefine man, or accept chimpanzees as human."

After some scrambling among anthropologists to redefine the uniqueness of our species, they landed upon our ability to use tools *to make* other tools. David Greybeard may strip away the branches of his termite dipping-stick, but only hominins (a catchall word that refers to *H. sapiens* and all of our extinct ancestors after the split from apes) could invent a special branch-stripping tool. Many archaeologists I spoke with believe the ability to plan and solve a problem using a complex device does not merely define our species, but in a few instances made our species. Our inventions aren't the result of our evolution, they believe, but are instead the explanation for the route it took. In at least a few cases, the earliest first inventors did not merely enable a new way of life or allow new economic possibilities, as we would modernly think of a modern invention, but instead enabled our evolution.

In no case is this truer than in that of the very first invention of all, made by an ancient ancestor of ours long before *H. sapiens* evolved.

Who was the first inventor?

I'll call her Ma, because she was a young mother, who like all inventors, had a problem.

Ma was born approximately 3 million years ago and belonged to an ancient ancestor species of ours called *Australopithecus.* She was born in Africa, perhaps Eastern Africa, where archaeologists have discovered a concentration of australopithecine fossils, including the famous "Lucy" found in 1974. Three million years ago is approximately halfway from the time when our species first split from the chimpanzee and bonobo line to the modern day, so it's no surprise that in appearance and behavior, Ma represents a middle ground between *H. sapiens* and chimpanzees.

She stood almost four feet tall, weighed a lithe sixty-five pounds, and other than on her hairless face she was covered in thick dark fur. Ma ate more meat than a modern chimp does, but she scavenged it rather than killed it. Ma supplemented her meals with roots, tubers, nuts, and fruits. In many respects, a modern observer might mistake her for a remarkably well-balanced, walking chimp, save for her peculiar, dexterous, and inventive use of rocks. To aid her work scavenging carcasses, Ma sharpened stones to cut into bones for marrow, which allowed her to eat meat other scavengers couldn't access.

Ma was a clever ape, but to many of Africa's big cats, she was still lunch. During the day she walked upright in search of food, but at night she clambered back into a tree nest to avoid nocturnal predators. Archaeologists have found australopithecine femurs and arm bones in caves adjacent to complete predator skeletons, which is a clear but grim signal of who was eating whom.

The predators interested in Ma were varied. She lacked fire and as a result found herself particularly vulnerable to a hunter similar to the modern panther, but she occupied a rung so low on the food chain that even eagles made the occasional meal out of australopithecines.

Her inability to start and control fire had a far more significant implication: It meant she ate her food raw.

The digestive system extracts fewer calories from raw food than cooked and it is far more difficult to chew, which means

Ma had to spend more time gathering and eating than a modern *H. sapiens*. Even with their large teeth and strong jaws, modern chimpanzees spend up to six hours per day chewing their raw food, while the average modern person's cooked diet allows them to eat a day's rations in a brisk forty-five minutes. Ma's raw diet meant she would have had to spend nearly her entire day gathering food and eating it while dodging eagles and panthers, clambering up and down trees, and roaming across open ground looking for carcasses and fruit.

All of which would have become far more difficult when, in her early teens, Ma gave birth to a noisy, helpless, immobile infant.

H. sapiens infants are an evolutionary curiosity. Most mammalian babies are born ready to walk, trot, or at least hold on to their moms. The reason is blindingly obvious: Every day a baby spends unable to keep up is life threatening for both mother and child. A capuchin monkey's baby can grip its mother's fur almost immediately, while the bigger-brained chimpanzee's mother has to carry her newborn, but only for its first two months. *H. sapiens* babies, on the other hand, spend more than a year in almost complete helplessness, unable to walk, crawl, or even support their own body weight. While this would seem an evolutionary disaster, it is the downside to what is perhaps our greatest strength: oversize brains. Our extended weakened state is partially explained by the time required to develop trillions of synaptic connections within our brain. In all primates, an evolutionary trade-off occurs between larger brains and infant mortality, and each species has arrived at its own equilibrium. The question archaeologists have asked is how humans arrive at such a perverted one.

Presumably, when hominins first branched off the chimpanzee line, hominin babies could soon cling to their mothers. Yet at some point this began to change. When I asked Cara Wall-Scheffler, a biologist at Seattle Pacific University, when young hominin mothers would first have been strained to the breaking point by their helpless babies, she said she believes the switch to bipedalism nearly 3 million years ago would have placed mothers and their newborns in a dangerous position.

Her reasoning is straightforward: Walking upright would have made it far more difficult for a baby to cling onto its mother. In addition, upright walking requires narrow hips, which would have narrowed the birth canal and necessitated smaller-headed babies. But instead of hominin heads shrinking, and hominin babies becoming more capable, the exact opposite occurred. Head size increased, and babies became even weaker. Today, *H. sapiens* have one of the largest body-to-head-size ratios in the animal kingdom despite walking upright. It's an oddity biologists call the smart biped paradox.

The evolutionary explanation for the paradox is that hominin mothers like Ma birthed their babies earlier in their gestation. Essentially, Ma's baby was born two or three months premature, before its head could outgrow the exit. Since Ma, the change has only become more pronounced. If *H. sapiens* birthed babies at the same developmental stage as chimpanzees, pregnancy would last twenty months. Not only would a baby that large not fit the birth canal, but the strain on the pregnant mother would be far too great. The result is that a human baby's first seven months are spent as if it were still in the womb—helpless and completely dependent on its mother—while the baby adds more than a billion synapses to its brain every minute.

Ma's helpless baby would have posed the greatest challenge to her while she gathered food. No modern primate species, with the exception of ours, shares parenting duty, so she is unlikely to have received help from the father. Nor is she likely to have even set her child down for longer than a few moments, because no primate in the wild *ever* parks their baby. It's simply too dangerous. If Ma left her baby while she gathered food, her baby's reaction would have been quite like what you would hear from a human baby in a similar situation today. Eventually, her baby probably wouldn't have been there when she returned.

The cumulative evidence suggests that Ma would have had to carry her baby for at least its first six months of life while spending most of her waking hours searching for food. The exertion of energy alone would have been life threatening. Wall-Scheffler looked at the ergonomics of the baby-carrying problem an australopithecine mother like Ma would have faced, and concluded she would have expended 25 percent more energy than usual while carrying her baby—far exceeding the already significant cost of nursing. In her estimation, carrying a baby is so taxing, bipedalism itself would have necessitated a solution.

Ma's solution, Wall-Scheffler tells me, was the astoundingly revolutionary, species-altering idea to not only invent something, but invent what is probably the most consequential tool of all time.

She invented a baby sling.

The materials of Ma's sling would have been basic. Perhaps as simple as a single loop of vine wrapped and tied off in a knot. While a knot-tying may seem advanced for an australopith like Ma, all great apes can tie knots, and as Wall-Scheffler told me, it therefore "doesn't seem outside the realm of possibility that australopiths would have been able to make a simple loop."

The baby sling therefore may not have been as much of a technical challenge as a conceptual one. Using a tool to make another one involves what psychologists call "working memory," which simply means the ability to hold information in your mind, manipulate it, and then use it.

Working memory is something we employ all the time—for example, when you shop at the grocery store, you might visualize the dish you're going to make so you can buy the necessary ingredients. Or if you're completing a puzzle, you might visualize what it's supposed to look like so you know where a given piece should go. The more steps that are involved in a given task, the more working memory is required. Building a rocket part that has a complicated interplay with thousands of other components requires more brain power than shopping for dinner, but the principle is the same. Ma couldn't build a rocket, but when her baby weighed heavily upon her and she visualized a potential solution, she demonstrated the early beginnings of a sophisticated psychological trick.

Ma's sling may have made her life only slightly easier, but its evolutionary consequences are difficult to overstate. A simple sling would have allowed hominin babies to spend a nearly unlimited amount of time in a helpless state, which, according to archaeologist and author of *The Artificial Ape*, Timothy Taylor, didn't just alter the smart biped paradox, it threw it out entirely. The paradox doesn't exist if mothers, armed with slings, can birth their babies long before they would otherwise be developmentally ready. A baby sling didn't just ease Ma's burden. It removed the evolutionary governor on how large our brains could grow. In doing so, the baby sling altered our evolution.

It sounds hyperbolic. It isn't. Without baby carriers, helpless hominin babies would have been set down by tired mothers and picked off by panthers long ago. According to Goodall, inexperienced chimpanzee mothers lose half of their babies while they're unable to hang on. And that's in only two months. Large heads on bipedal creatures should be an evolutionary dead end. The fact that it isn't is thanks to the baby sling. And to Ma.

Of course, if Ma had been the only one to use a baby sling, and if her fellow australopithecine mothers had given her invention no more than a quizzical look, the evolutionary consequences would have been nil. She would have made her life slightly easier and nothing more.

But that's not what happened.

Ma's invention spread. Shortly after Ma, according to Taylor, our ancestors experienced a burst of rapid brain growth. This dramatic growth, which resulted in mothers birthing babies even earlier in their development, would have been impossible without her sling. And if Ma's idea spread, then it suggests that australo-pithecines already possessed the beginnings of what might be *H. sapiens'* greatest skill: We are a species of incredible copiers.

Anthropologists call this skill "social learning," and when researchers have tested *H. sapiens* newborns against chimps in a variety of intellectual challenges, social learning is the skill at which humans exhibit far more talent. According to the Harvard University professor of human evolutionary biology Joseph Henrich, *H. sapiens* are habitual mimics. We watch each other, we learn, and we copy. Essentially, we are a species of shameless intellectual plagiarists. But this is a feature, not a bug.

None of us are as terribly ingenious as we might like to think, particularly when it comes to our survival. As Henrich notes in his book *The Secret of Our Success,* stranded groups of human explorers who have found themselves shipwrecked or abandoned in the deserts of Australia or on the frigid tundra of Greenland have a wretched record. In nearly every case, explorers lost in novel environments either accept help from locals or starve to death in their ignorance.

The lesson, according to Henrich, is that we owe our incredible adaptability to humanity's ability to learn, copy, and compound small innovations. If humans, like apes, largely ignored each other's moments of inspiration, we might still be stuck in the same environmental niche. But thanks to our relentless excellence in mimicry, we aren't. Each small improvement by each individual is monitored, learned, and adopted by the group. We are the technological ratchet machines of the animal kingdom. Through micro innovations and collective plagiarism, we progress.

By the time of Ma's invention, hominins' great mimicking ability seems to have already existed, because her fellow australopithecines did not ignore or mock her curious contraption. They did what hominins do best: They copied it.

Not only did the wide adoption of the sling artificially extend gestation and remove the upper limits for how large our brains could become, it tightened the bond between mother and child. The sling physically attached a mother and child in such a way that they could look at each other for long periods of time. And while Ma did not speak a complete language, she could almost certainly, like chimpanzees, communicate on a simple level.

Goodall has recorded communication between chimp mothers and their babies, and their calls largely consist of vocalized "hoos" when a baby wants a ride or the mother wants them to climb aboard. While this is nothing like the constant chatter that occurs between a *H. sapiens* mother and her newborn, the increasingly frequent hoos Ma cooed at her newborn may have been a precursor to something much more sophisticated. "Motherese"—the melodic way in which a mother speaks with her infant, also known as baby talk, e.g., "Aren't YOU a GOod GIrl"—is a cultural universal. All human mothers of all languages speak to their babies in the same rhythmic HIGH-low-HIGH pattern, suggesting to anthropologists that this way of communicating stretches deep into our past. Some linguists even believe baby talk is an echo of the original tongue, developed long before language evolved. As Ma knotted the first sling and spent her days looking into the eyes of her infant, she may have unintentionally strengthened the critical bond between mother and child forever.

This bond may have led to language, increased socialization, greater intelligence, and far more sophisticated inventions, but all of those developments took thousands of years. Ma may have eased her burden and improved the likelihood of her child's survival, but there's no evidence Ma or any australopithecines evolved these more sophisticated social relationships. Nor is there evidence they honored or remembered their dead, even if the dead was their own mother. So when Ma passed, no one eulogized or even buried the world's first inventor. Instead, her child may have simply carried her body away.

2

Who Discovered Fire?

This occurred 1.9 million years ago, which is
before humans evolved.

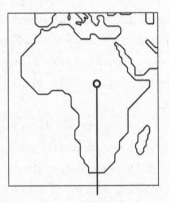

1.9 million years ago

Control of fire

Between 1891 and 1892 on the island of Java, Indonesia, a
Dutch physician-turned-paleoanthropologist by the name of
Eugène Dubois discovered a thigh bone, molar, and skullcap be-
longing to an ancient individual with a mysterious near-human
shape. Dubois famously declared that Java man, as he was eventually dubbed, was the "missing link" between humans and apes.
The academic community was considerably less enthused.

They rejected his theory out of hand, and proceeded to engage in a lively debate over the degree to which Dubois deserved ridicule. The naturalist Richard Lydekker wrote that the skull was likely that of a deformed human, while the German anatomist Wilhelm Krause declared that Dubois had indeed made a great discovery—but only because he had identified a new species of gibbon.

It wasn't until 1921, when the archaeologists Johan Andersson and Walter Granger uncovered the nearly identical bones of forty other individuals in a cave system outside Beijing, that the scientific community recognized Dubois' discovery for what it was: a staggeringly significant confirmation of Charles Darwin's theory of man's evolution.

Biologists called this new species *Homo erectus* (upright man) and for a time believed it was indeed the missing link—the single species that connected *Homo sapiens* to the great apes. But forty years later, in east Africa, Mary and Louis Leakey discovered evidence of another, older primate species. They called it *Homo habilis* ("handy man") for the assortment of chipped rock tools they found nearby. The question then became how the two species were related.

At first it wasn't obvious that they were at all, because the differences between the two species are stark. Java man hunted, slept on the ground, had little body hair, a far larger braincase, a far smaller jaw, and a dramatically smaller chest, while handy man scavenged, walked the ground in the day but slept in tree nests at night, and was covered in fur. Many biologists would prefer to remove *H. habilis* from the *Homo* genus entirely, while *H. erectus* is so humanlike that Lydekker mistook it for one.

Yet over the course of an incredibly short amount of evolutionary time, many scientists began to suspect a group of *H. habilis* did, in fact, evolve into *H. erectus.*

What could account for this radical transformation? According to Harvard primatologist Richard Wrangham, the entire suite of morphological changes can be explained by one simple yet profoundly consequential, species-altering trick: Someone learned to control fire.

For all their differences, *H. habilis* and *H. erectus* did share one thing in common: They chipped stones to make sharp cutting tools. This ancient practice represents the oldest manufacturing process ever devised. It led directly to knives, arrowheads, and eventually Saturn V rockets. But it also led to the far more consequential discovery of fire.

Charles Darwin called the mastery of fire the single greatest discovery in the history of our species, excepting language. If anything, that's an underestimation. Fire wasn't controlled by our species; it's responsible for it. Fire didn't simply make life easier—it forged us, principally and most dramatically, through cooking.

On the most basic level, applying heat to food outsources chewing and digestion. It not only softens food, but breaks down its chemical bonds. This sounds like trivial convenience. It isn't. The smaller and softer the pieces of muscle, fat, sinew, and cellulose in food—and the more their chemical bonds are broken—the more of their energy the intestines can absorb. The result is that food, either from plant or animal, provides 25 to 50 percent more calories when it's cooked as opposed to when it's eaten raw.

The control of fire therefore resulted in a massive caloric influx to which our bodies have long since adapted. *Homo sapiens* have evolved to eat cooked food as much as the giraffe has evolved to eat the highest leaves. By now our jaws are too weak, our teeth too minuscule, our stomachs too small, our intestines too short, and our brains too calorically glutinous to go back to a fireless world. There is no evidence of any modern *H. sapiens* ever surviving on a diet of wild and unprocessed raw food for more than a few weeks. Our species would be at risk of extinction without cooked food.

The mastery of fire was not the result of the enormous behavioral and physical changes between *H. erectus* and its more apelike ancestors. It caused them. No invention, discovery, or insight made before, since, or possibly in the future could surpass the skeleton-shifting influence the application of heat to food had on our species. And unlike the development of language, the mastery of fire was not an evolution. It was a discovery.

Who made it?

I'll call her Martine, after the seventeenth-century French geologist Martine Bertereau, both because the mastery of fire is primarily a geologic discovery and because Martine Bertereau was jailed for witchcraft, which you can imagine is an accusation our Martine, after striking the first fire, would almost certainly have risked as well.

Martine was a *H. habilis* born approximately 1.9 million years ago in East Africa, a long, long time before the emergence of anatomically modern humans. She stood approximately four feet tall, weighed seventy pounds, and had a braincase roughly 40 percent the size of a modern *H. sapiens*. Her forehead sloped gently to jaws that jutted forward to house a set of larger teeth that had a far more powerful bite than a *H. sapiens*. Her skeleton indicates she occupied a curious halfway point in hominin transition from living in trees to walking upright. Her legs and hips were well adapted for walking, while her arms were elongated and her shoulder structure retained adaptations for climbing. Archaeologists believe Martine was largely bipedal, spending her days searching the African savannah for nuts, berries, and scavenged carcasses. But she also slept in trees—likely to avoid nocturnal predators.

Martine occupied a middle rung of the food chain, and archaeologists regularly find *H. habilis* fossils with claw and tooth marks in their skeletons. She was not yet the endurance hunter that *H. erectus* became, as her legs and Achilles tendons are both far too short to give her the efficient stride of *H. erectus*. Besides, she was covered in fur and would have quickly overheated on a long run.

She did eat meat, but likely scavenged it or killed opportunistically, as chimps do today. Unlike *H. erectus*, who set out on long endurance hunts, Martine subsisted as a gatherer and a scavenger. The proof comes not only from her bone composition but also from a DNA analysis performed on a modern-day parasite called the Taenia tapeworm. This worm, DNA suggests, originated in hyenas but was passed on to our ancestors some 2 million years ago—about the same time as when Martine lived—when an unfortunate hominin ate an antelope carcass previously dined upon by an infected hyena.

Scavenged meat is difficult to butcher without the flesh-ripping teeth, beaks, or talons the animal kingdom's scavengers typically possess, so Martine compensated by sharpening stones. This doesn't seem like a great intellectual feat, but as the Stanford University professor of archaeology John Rick tells me, Martine would have to know exactly how the stone would break, where to hit it, how to hit it, and how to hold it. Chipping rocks in the precise way you want is more difficult than one might imagine, and without watching someone else do it first I doubt many could figure it out.

In chipping her cutting stones, Martine demonstrated an even more important skill with regard to her ability to start a fire: A good stone-tool maker would also need to be able to differentiate between various types of rocks. She would have selected a hard, brittle rock such as flint or obsidian and struck it with a solid river stone to make her cutting tools.

Over time, after chipping hundreds of thousands of stone tools, *H. habilis* like Martine must have sparked the occasional fire, writes Wrangham. That in itself would not be terribly groundbreaking.

Fire is a familiar phenomenon to the chimpanzees on the African savannah, and Martine would already have been acquainted with its range of devastating and useful effects. Chimps in Senegal are known to successfully navigate around their habitat's regularly occurring wildfires. They can predict a wildfire's path and sometimes even seek it out in order to forage the scorched grasslands for cooked food, which is how all mammals, including chimps, prefer their meals.

The preference for cooked food is deeply instinctual. A baked potato tastes better than a raw one because natural selection has tuned our taste buds to prefer food that will provide more calories. Cooking a potato doesn't add calories to it, but it increases what our intestines can extract from it and we therefore prefer a baked potato over a raw one.

This doesn't just apply to *H. sapiens*, either. Every mammal's digestive system can extract more calories from cooked food, which explains why not only humans but chimps and even rats prefer their potato cooked if given the opportunity. Some scholars believe domesticated dogs evolved from wolves that foraged through early human trash pits for scraps of cooked food. Eventually their digestive systems, like ours, evolved around this new diet. This universal preference for cooked food suggests early hominins, like chimps, sought out wildfire and perhaps hoarded it opportunistically.

But opportunistic fire is not ever-present fire, and the *H. erectus* who journeyed from East Africa to Indonesia already possessed an intestinal structure that required cooked food. This suggests they not only hoarded fire but knew how to light a new one when the previous one was inevitably extinguished. Someone must have learned to control fire.

Martine's stroke of genius would not have been the moment she sparked a fire, which would have likely been an unusual but not unique occurrence. Instead, her genius would have been her insight into why rock-striking occasionally sparked fires *but usually didn't.* The answer comes down to geology. Lighting fires by chipping stones is either straightforward or impossible, depending on which stones you choose. The key is pyrite.

Pyrite is an iron sulfide, and it presents in different forms. It's sometimes called fool's gold for its visual similarity to the real thing. Fool's gold is a poor fire starter, but when I attempted chipping a particular type of pyrite called marcasite, I was able to spark a fire within a few minutes.

Hominins have used pyrite to ignite fires for as long as the archaeological record can indicate. The five-thousand-year-old Ötzi the Iceman had a fire-starting kit of pyrite flakes, flint, and fungus tinder inside his pack. In Belgium, archaeologists found a thirteen-thousand-year-old grooved pyrite block that was clearly used to spark fires. And in 2018, Leiden University archaeologist Andrew Sorensen discovered pyrite residue on fifty-thousand-year-old Neanderthal hand axes—suggesting Neanderthals lit fires by knocking pyrite against the side of their stone tools.

Pyrite is common across the globe, including in East Africa. A modern pyrite, gold, and silver mine in northern Ethiopia is only four hundred miles from some of the oldest *H. erectus* skeletons in Koobi Fora, Kenya. The use of pyrite—or, more modernly, steel—to start fires is essential. Other stones simply won't work because they're missing the essential ingredient: unexposed iron.

When iron is exposed to oxygen, it combusts. In large quantities this process occurs slowly and presents as the familiar rust. But in small chips with low volume and high surface area, the iron in pyrite oxidizes so quickly it generates enough heat to ignite tinder.

Marcasite is a dark, brittle stone that could have been easy to mistake for a river pounding stone, though once it's split open the unexposed iron sulfides shine like gold. If Martine happened to sharpen her flint or obsidian tool against marcasite, she would have showered the ground with small pieces of unexposed iron. And if she chipped her tool over dry grass, she could have started a fire entirely by accident. This would have been an alarming if not terribly unique occurrence. The sheer number of chipped stone tools would have made it almost inevitable. But Martine's insight that the dark, brittle rock with shiny flakes was the key is unquestionably the most consequential observation in hominin history.

It might seem beyond the intellectual capacity of a hominin with a brain less than half the size of ours to make this geologic insight, but Martine already differentiated between stones to craft her tools. And it may seem beyond the ability of this distant ancestor of ours to start a fire, maintain it, and then—crucially—realize why it all happened. But in 2005 when the trained male bonobo Kanzi was given only matches and a marshmallow by the primatologist Sue Savage-Rumbaugh, Kanzi could gather kindling, spark a fire, and toast his marshmallow. Martine had a brain twice the size of Kanzi's.

Fire changed virtually every aspect of hominin existence. Sleeping on the ground without fire in the presence of Africa's great predators is dangerous even for modern hunter-gatherers.

Yet the skeletal evidence is unequivocal: *Homo erectus* slept on the ground. Without fire, it would be difficult to explain how. With her fire, Martine left her nest to sleep by it—and as a result, hominins lost the adaptations for climbing trees.

Sleeping near a fire also outsourced fur's primary purpose: warmth at night. Freed from the need for thick fur, hominins with thinner hair (*H. sapiens* have the same number of hair follicles per square inch of skin as chimps—our hair is just thinner) became more efficient hunters than those with the full complement. *Homo erectus* could cool itself more efficiently than the animals they chased, and if they became cold at night they simply moved closer to the fire.

Sitting around fires also made people nicer. No one knows how effectively Martine could communicate with her fellow *H. habilis*, but whatever ability she did have, fire almost certainly strengthened it. Sitting in close proximity to a campfire for its warmth and food forced Martine and her band to cooperate. A violent, uncooperative, or unstable individual would risk banishment from the fire and a virtual death sentence. Gradually, trustworthiness and predictability yielded survival value, and evolution began to favor brains that could better navigate social relationships born around a campfire. In essence, the use of fire favored individuals with larger, nicer brains.

Large brains are calorically expensive—our brains consume 20 percent of our energy—but here too fire helped. Not only did fire increase the number of calories Martine could extract from her food, but by obviating the need for her strong jaws, huge stomach, and long intestines, fire eventually resulted in a radical reallocation of the body's resources.

In "The Expensive-Tissue Hypothesis," the paleoanthropologists Leslie Aiello and Peter Wheeler argue that *H. erectus* partly paid for its 30 percent larger brain by ridding itself of the previously robust and now superfluous parts of its digestive system. These radical changes are difficult to explain if not for a full-time transition to cooked meals.

Then there is the matter of the time cooking saved, which seems like an oxymoron, but if cooking takes time, chewing raw, inefficient food takes far longer. As we've seen, the amount of time chimps spend eating is more than six times that of modern humans, a dramatic difference explained by both the toughness of raw food and the fewer calories extracted by each bite. Because *H. erectus* was larger than a modern chimp, Aiello and Wheeler estimate a *H. erectus* on a raw food diet would have had to eat for eight hours per day to maintain their body weight. By cooking, *H. erectus* added hours to their day, and the evidence suggests they used it to pursue the foods they valued.

One of the many consequences of cooking was a dramatic increase in the time *H. erectus* spent hunting. Chimps only eat meat when they stumble upon prey, according to Wrangham, and never embark upon hunts, because in the likely event they're unsuccessful, they will starve. The same was likely true for Martine, who was more often the hunted than the hunter.

Yet *H. erectus* was a hunter. *Homo erectus* sites routinely exhibit scattered bones from their meals. And they could hunt, according to Wrangham, because when a *H. erectus* arrived home in the evening after an unsuccessful hunt, he or she could quickly eat a cooked meal.

Of course, Martine would not have realized any of these evolutionary benefits, which took tens of thousands of years to evolve. Martine lived out her life as a *H. habilis*—though perhaps a happier, heftier one considering her heartier meals. Fed and protected by her fire, she may have even enjoyed the rare privilege of dying of old age.

Upon her death, you would like to think that her community honored this hominin for her discovery. Perhaps, fittingly, she could have been cremated. Unfortunately, *H. habilis* didn't honor their dead, so the only ceremony her companions would have likely bestowed upon the body of the greatest hominin who ever lived would have been to haul it away so it didn't attract scavengers to the campfire.

3

Who Ate the First Oyster?

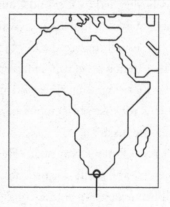

This and all subsequent events occurred after humans evolved. If the time of our species on earth were a day, **this happened at 10:53 in the morning** (164,000 years ago).

164,000 years ago

First eaten oyster

In the summer of 2007, a team of archaeologists, led by Curtis Marean of Arizona State University, working in a set of caves on the southern cape of South Africa, found fossil evidence of a 164,000-year-old *Homo sapiens* encampment. They uncovered the charcoal remains of ancient campfires, stone tools, red pigment, and, most critically, shucked oysters—the oldest evidence that someone had braved eating the slimy oyster. But according to Marean, the shells signify far more than an adventuresome palate. He believes these crustaceans were the edible reward for a difficult and obscure astronomical observation made by a scientist who lived more than 100,000 years ago.

Who was this scientific hero and culinary daredevil who braved the first oyster?

Her name was Oyster Gal—or at least that's what I'll call her. The name isn't as far-fetched as it sounds, and it may even be a rough translation of her actual name. Naming conventions throughout history are usually descriptive (see "Johnson," "Smith," "Baker," et al.), rather than a set of random, pleasing sounds. And if her name did have meaning, what else could anyone call the first person to crack open a rocklike object found in a tide pool and eat the pale, snotty creature inside?

Jonathan Swift once famously mused, "He was a bold man that first ate an oyster," and Swift was probably right that eating a pale lobe of shellfish meat was a courageous choice. But he was probably wrong about the other part. The evidence suggests it wasn't a bold man that first tried the oyster; it was a bold woman.

In every hunter-gatherer tribe studied by modern anthropologists there exists a gender divide for obtaining food. Regardless of whether the bulk of the work falls to the hunter or the gatherer, the division is remarkably strict: Women usually gather the staples that are more reliably obtained—such as nuts, berries, roots, and shellfish—while men typically pursue the food that runs, flies, or swims away. Even in the tropical islands in the north of Australia, where gathered foods were plentiful, the men did little food gathering. And in Tierra del Fuego, at the southern tip of South America, where the majority of the calories came from hunting large sea mammals, the women did not hunt.

There is no anthropological consensus that explains the gender divide in hunters and gatherers. There are theories, however. Richard Wrangham traces it to the control of fire, and specifically the time required to cook. Chimpanzees eat food as quickly as they can, lest a larger chimp steal it. Assuming early hominins behaved similarly, cooking would have been an impossibility for all but the largest. As a result, Wrangham theorizes, in order to cook females were forced into a kind of mafia-like protection racket with larger males who ensured a smaller female could cook in exchange for some of the food. With males now guaranteed a night's meal, they were free to pursue higher-value foods without risking hunger in the likely case of their failure. It's only a theory, and one that's nearly impossible to prove, but it would explain the curious male and female food partnership in *H. sapiens* that's unobserved in any other primate. It would also explain the strict sexual division of labor in hunter-gatherers, why staples like oysters are more likely harvested by women, and why a woman is more likely to have eaten the first.

Oyster Gal was a *H. sapiens*, which means if she were to sit on the bus next to you today, you would not be immediately alarmed. Her stature, body, face, and hair would look familiar. She had the same size cranium, jaws, teeth, pelvis, feet, and hands as we do. She would be shorter than modern standards, but her posture and gait would be entirely modern. Her skin would be hairless and dark, evolved to better shield against the intensity of the African sun. The hair on her head would be dark, short, and curly.

Her clothing, however, would be a little out of the ordinary. She probably didn't wear any, or at least nothing we would recognize as such. Sewing was unknown. The bone needle didn't appear for another hundred thousand years, and researchers now believe the first use of both clothing and shoes occurred remarkably late in our evolution. According to the anthropologist Erik Trinkaus, who has studied the toe bones of ancient *H. sapiens*, Oyster Gal's toes were considerably stronger than yours or mine, probably because she walked without supportive shoes. Trinkaus believes *H. sapiens* toe bones have atrophied in the confines of sturdy shoes, explaining the notable difference in the size of ancient toes. According to Trinkaus, "modern" toes don't appear until around the same time as the bone needle, suggesting *H. sapiens* didn't invent the shoe until long after Oyster Gal.

As an adult, Oyster Gal was likely the mother to children, though perhaps not many. In a study of the !Kung San hunter-gatherers of Africa's Kalahari Desert by the anthropologist Richard Borshay Lee, he found that the San women went a minimum of four years between children. Because they didn't use any formal birth control, however, the explanation for these long periods of infertility is unclear. One hypothesis, proposed by

Rose Frisch and Janet McArthur, is that !Kung women, while healthy, simply didn't have sufficient body fat to simultaneously nurse and menstruate. Nursing, according to their theory, may have provided Oyster Gal a natural form of birth control.

If she had a young child she would have communicated with it, and with her peers, but to what degree is a matter of great debate among archaeologists. Richard Klein, a professor of anthropology at Stanford and the author of *The Dawn of Human Culture*, tells me he believes *H. sapiens* of Oyster Gal's era probably lacked a modern language, citing their dearth of sophisticated culture or technology in their encampments. He dates the origin of modern language to a time long after Oyster Gal—approximately forty-five thousand years ago—when *H. sapiens* encampments exhibit an explosion in cultural artifacts. Sites dated to this era exhibit unequivocal evidence of modern behavior, such as art, music, and a belief in the supernatural.

Marean, on the other hand, believes Oyster Gal dreamed, laughed, and communicated fluidly and thoroughly. "The languages spoken at this time," he wrote to me, "were as rich as our own." As evidence, Marean cites the decorated shells and the natural painting pigment he has found in the Pinnacle Point caves in South Africa, both of which represent examples of the kind of symbolic thinking that archaeologists look for when determining an ancient culture's cognitive modernity. Marean also believes many of the tools she made were so complicated that the techniques to make them would have been impossible to teach and pass on without language.

Until recently, Marean's view was in the minority and most

archaeologists agreed with Klein. Oyster Gal and her contemporaries, they believed, differed in some significant cognitive way from modern *H. sapiens*. Now, with the discovery of the artifacts in the Pinnacle Point caves and other ancient examples of symbolism, scholars are far more split on how Oyster Gal viewed her world and what happened when she spoke.

She could certainly have communicated, but to what degree remains a matter of scholarly debate. However, there's no question she and her people lacked the significant technological innovations of later *H. sapiens*. There's no evidence, for example, that she used fishing poles, nets, or boats, which explains why *H. sapiens* encampments from Oyster Gal's era are rarely found along the coast. For Oyster Gal, the ocean represented a food desert, and as a result living on the beach would have needlessly cut her foraging area in half. This explains, according to Marean, why in the eras prior to Oyster Gal there's little evidence hominins ate food from the sea and no signs of shucked shells. A key reason, according to Marean, is the tides. Oysters are only revealed by the tide's lowest ebb, which means 95 percent of the time they would have been inaccessible. This infrequency and unpredictability probably explain why early hominins didn't gather them. According to Marean, "hunter-gatherers never base a subsistence system on gathering methods they cannot schedule or do not understand."

The Pinnacle Point caves contain thick deposits of oyster shells. Clearly, oysters had become a staple part of the diet, perhaps because Oyster Gal discovered the sign that foretold their appearance.

When Oyster Gal ate her first oyster she had likely ventured to the ocean in search of something else—a sleeping turtle or its eggs, or maybe a beached whale or resting sea lion. Archaeologists have discovered a type of barnacle among the artifacts in the caves called a *Coronula diadema* that lives only on the skin of humpback whales—a suspicious item in a cave three miles from the ocean. Eventually, on one of these trips, Oyster Gal would have by chance arrived at a super low tide and found a few exposed oysters. And then cracked one open and braved its meat.

With all due respect to Jonathan Swift, this might not have been quite as courageous as it appears.

A few other animals, including modern baboons, eat oysters, which if observed by Oyster Gal would have given her some confidence in their edibility. But baboons also eat blossoms and bark, and an experienced food gatherer like Oyster Gal would have been cautious about following the dietary example of other animals. It's an experiment that can sometimes end poorly. If she watched a rabbit eat a few berries from the *Atropa belladonna* plant, for example, and did the same, she would have died within the day.

Still, if she saw another animal eat an oyster, it might have provided the necessary courage. Cooking her oyster would have provided more. Cooked food is safer than raw—and caution is something the slimy, wrinkly oyster would certainly inspire.

Dietary boldness, however, was not Oyster Gal's real genius. Based upon how many oyster shells have been found in the Pinnacle Point caves, Marean believes Oyster Gal knew when to travel to the ocean.

In other words, she learned how to predict the ocean's tides.

Super low tides on the southern tip of the African coast occur for a few hours during a few days every month. But they aren't just infrequent, they're inconsistent. The level and timing of both low and high tide change dramatically throughout the month, from a slight difference between low and high tide to a large one, and the explanation remained a mystery to hominins for millions of years. It's easy to see why. The solution to their pattern lies in an incredibly unlikely and seemingly unrelated location: the night sky.

The ocean's tides are caused largely by the pull of the sun's and moon's gravity. When they both align—either on the same side of the planet or opposing sides—they pull together at earth's center and create a tide that is both super high and super low. Fortunately for Oyster Gal, this random-seeming event is advertised by the appearance of either a full or a new moon. Which means Oyster Gal wasn't just an adventurous diner. She was a stargazer.

This link between the movement of the ocean and the various shapes of a mysteriously large white object in the night sky is so incredibly nonobvious that many people today still don't see it. Oyster Gal, however, did, and in doing so may well have become the world's first practical astronomer. Once she could predict the tides, Oyster Gal could confidently schedule trips to the ocean, and the evidence from the caves suggests oysters became a reliable staple of her diet.

If her discovery of the moon's link to the tides was celebrated,

Oyster Gal didn't have long to enjoy her renown. As a twenty-something woman of ancient times, contemporary actuary tables suggest she was nearing the end of her lifespan. According to an analysis of tooth-wear done by Rachel Caspari, an anthropologist from Central Michigan University, two thirds of *H. sapiens* from this era who lived beyond childhood died in their twenties and very few made it past thirty-five. Although it's impossible to know exactly how she died, we can say with some confidence how she didn't. Nearly all of the leading causes of death today were prehistorically unknown, either because they are primarily the diseases of old age (cancer, heart disease, stroke), or require large, concentrated populations to support (cholera, typhoid, and flu). Neither of those circumstances existed during Oyster Gal's era.

The far more likely causes were childbirth, malaria, accident, homicide, or bacterial infection. She may have been attacked by a predator while foraging. Or, perhaps her adventuresome palate caught up with her.

Once she passed, she may have been buried by her people. But whether her burial was to honor her memory or simply to discourage scavengers is again the subject of contentious debate among scholars. Archaeologists have found bodies older than Oyster Gal buried, but the graves don't show any of the gifts or signs of care that would eventually become universal across all cultures.

The first evidence of a ritual burial comes from a grave in Israel, where archaeologists have found the skeleton of a man with the jawbone of a boar on his chest, dated sixty thousand years after Oyster Gal. As a result, it's impossible to say whether Oyster Gal, a prehistoric scientific genius who helped solve one of the greatest problems of her time, had a memorial service. If she did, hopefully an oyster shell was placed on her chest. Ideally one that was in the shape of a crescent moon.

4

Who Invented Clothing?

If the time of our species on earth were a day, **this happened at 2:34 in the afternoon** (107,000 years ago).

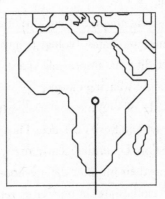

(Somewhere in Africa)

107,000 years ago

First clothing

When our hominin ancestors first branched off the chimpanzee and bonobo line, they hosted but a single species of bloodsucking, fur-living parasite: *Pediculus humanus capitis*, also known as the head louse. It has lived on hominins for at least 6 million years and lives in our hair to this day. But the head louse isn't the only species of louse we host. Over the past 6 million years, we have brought aboard two more species of bloodsucking parasites: the pubic louse and the body louse.

We picked up the pubic louse approximately 3 million years ago when an ancestor of ours and a gorilla shared sleeping spaces (though probably not at the same time). Sometime later, the body louse, which clings to clothes instead of hair, emerged. The result is that today we host three times as many species of lice as any other primate.

However, this seedy chapter of our evolution has a silver lining. Thanks to these parasites, biologists have been able to answer what seemed like an impossible question: When, where, and why did we begin wearing clothes?

Clothing is made of organic material, which means unlike stone or bone it leaves no fossil evidence. Thus, its origin has long been an open question. Up until recently, many archaeologists believed hominins lost their fur after the invention of clothing made it unnecessary. As clothes are, in one sense, removable fur, scholars believed that at some point hominins upgraded from permanent warmth to warmth on demand. It's a sensible theory, but one that geneticists have recently discovered is spectacularly incorrect.

In 2002, David Reed, a curator of mammals at the Florida Museum of Natural History, analyzed the DNA of both the common head louse and the body louse and discovered that they did not diverge simultaneously from a common ancestor. Instead, the body louse evolved directly from the head louse. Presumably, he wrote, this occurred when *H. sapiens* first donned clothing and inadvertently provided the head louse with a new habitat. When Reed dated the divergence, he arrived at an astonishing result: Our ancestors spent a rather staggering amount of time naked.

According to Reed, the first person to wear clothes lived approximately 107,000 years ago. Yet the geneticist Alan Rogers has determined hominin skin color changed from a chimplike pale pink to UV-protecting darker hues some 1.2 million years ago—suggesting hominin body hair had by then thinned to an immodest degree.

How then did hominins survive for so long without fur or clothes? And, most important, who ended the million-year streak of nudity?

I'll call him Ralph, after Ralph Lauren, because the evidence suggests that when our Ralph made his insight, he was interested in fashion as much as function. (And I'll call him a him because in truth, I don't know. I flipped a coin.)

Ralph was a modern *H. sapiens*, born roughly 107,000 years ago in Africa, perhaps in South Africa. Many researchers believe that if Ralph were placed in a time machine, we could learn his language, and he could learn ours. They believe he had a modern mind, fully capable of symbolic thought and cognitively indistinguishable from a modern person. If he grew up in today's world, presumably he could be a fine banker, waiter, or lawyer.

Ralph is considered a modern *H. sapiens* in appearance, but he did have a rather unusual visage. When I asked Duke University archaeologist Steven Churchill if Ralph would have garnered any stares walking down a modern sidewalk, he says Ralph would have had a "somewhat more masculinized face" than that of a modern man. The primary difference, Churchill says, would have been a brow ridge that protruded to a degree rarely seen today. Brow ridges vary in size between person to person today, but even large modern ridges pause in their protrusion above the nose. Ralph's would have taken no such break, and extended forcefully for the width of his face. Endocrinologists believe this stern feature was the result of a stronger shot of testosterone than typically occurs in modern humans; thus a brow ridge is interpreted as a sign of aggression. Recent research suggests these ridges didn't recede until approximately eighty thousand years ago, when our ancestors began valuing an eyebrow's motions— and thus the ability to express dynamic temperaments—over permanent orneriness. Ralph would have had a stern and intimidating look, Churchill tells me. "But other than that, I don't think there's much that one might notice as different from people today."

By the time of his great invention, Ralph was likely an older, more influential member of his group. He spent his days hunting with spears, spear-throwers, and slings, though he was typically unsuccessful. He wielded a sophisticated rock tool set, which included hide scrapers, hand axes, and spear points, but he also manipulated organic materials such as vines, which he used for rope, and animal hides, which he used for bedding. He fought with other groups of *H. sapiens* and other archaic hominins over resources or food, or maybe just out of fear.

On his hunts, if he happened to successfully kill a warthog or deer, he shared it first with his kin and then among his group. If no one brought home meat that day, he ate tubers and roots and probably grumbled about it. He spent his evenings gossiping around campfires and at night he told stories about the stars. He smiled and laughed and played with his children.

And though he did all of this naked, he wasn't cold.

Cold is a challenge for *H. sapiens* because our physiological responses, namely shivering and rerouting blood supply, are, as the archaeologist Ian Gilligan puts it, "sluggish and ineffective," which is of course why most mammals are furry. The rabbit, for example, can withstand temperatures of as low as 49 degrees below zero. But shave its fur and that number rises to 32 above zero. This is why the million-year gap between the loss of fur and the use of clothing is so surprising. Other largely naked, land-based mammals either live underground or, like the elephant, have small surface to weight ratios that keep them warm. No other primate has as little fur as we do. How could warm-blooded, unclothed *H. sapiens* survive with no fur? The answer is the same reason hominins lost their fur in the first place: fire.

Hominins survived for the better part of a million years with neither clothes nor fur by lighting fires. Armed with fire, clothing becomes surprisingly inessential in all but the coldest of environments. Ethnographic observations of a few hunter-gatherer cultures prove it.

When the Portuguese explorer Ferdinand Magellan first sailed along the southern coast of South America, where the mountains are covered in year-round snow and the glaciers slide into the sea, his sailors discovered the people of the Yahgan and Alaculef tribes living quite comfortably without clothing. To stay warm, they coated themselves in animal grease and made heavy use of campfires—so much so that the sailors named it Tierra del Fuego ("Land of Fire"). It's not as if the idea of clothing never occurred to the Yahgan or Alaculef. They frequently warred with their clothed neighbors, the Ona people. They apparently just didn't find clothing necessary. The same is true for the aboriginals of Tasmania, who also made little use of clothing yet lived in far colder climes than Ralph.

Clothing is not what anthropologists refer to as a "human universal." It is not ubiquitous across all human cultures, as the control of fire is. However, decorating one's self—be it through painting, tattoos, earrings, or body modifications—is. Members of every human culture ever observed, from the uncontacted tribes of Papua New Guinea to Wall Street bankers, decorate themselves in some way. And because of this ubiquity, anthropologists assume humans today share this trait with all *H. sapiens* cultures. The Yahgan had little use for clothing, but they wore elaborate capes, bracelets, and necklaces and painted their faces and bodies. Individualization, anthropologists believe, is as basic a human need as food and shelter.

This deeply human truth suggests Ralph's inspiration probably did not come out of a need for warmth and certainly did not evolve from any modern Western ideals of modesty. Instead, it came from the same source of human desires that inspired the necklace, the tattoo, and the earring. In other words, Ralph's original piece of clothing was a fashion item.

Vanity is a powerful motivator for technological innovation, and Ralph is far from alone in inventing something that initially served little practical purpose other than to service his own. Anthropologists call inventions like these "prestige technology," because they function largely to bequeath status upon their users. Many of humanity's greatest inventions, including most metals, pottery, and fabrics originated as difficult-to-produce items that offered little practical advantage over alternatives. Yet they were in demand thanks largely to the difficulty of crafting them, which created a scarcity and thus a value. It was only later, through improvements in manufacturing, that products like copper or bronze gained a practical usefulness. Such may have been the case with Ralph and his clothing.

Ralph's invention may have started as something as impractical as a necktie, but it filled a deep human need, and over time it evolved into a critical piece of equipment that later *H. sapiens* used to survive in the extreme cold of the northern latitudes. By the time *H. sapiens* advanced into Europe fifty thousand years ago, they likely wore highly sophisticated furs and lined jackets. The bones in their archaeologic sites suggest they preferentially hunted wolverines, an animal that provides little meat but whose pelts would have been prized as insulation. The oldest image archaeologists have found of clothing is a twenty-four-thousand-year-old carving of a person wearing a close-hooded, highly practical parka.

But practicality came late to the invention of clothing. Whatever piece of clothing Ralph invented, it was closer to the bow tie than the shirt. Maybe he needed something to wear into battle when *H. sapiens* often take extra care to look their intimidating best. Or maybe he was leading an initiation or ceremony and required something to stand out. Perhaps his clothing was as simple as the shoulder-covering cape that the Yahgan occasionally wore. Or maybe it was something even less practical and more extravagant. The ethnographic record so abounds in hilariously impractical human clothing choices—from wigs to bow ties to phallic shields—it's folly to guess what Ralph might have come up with.

We do know that whatever it was, it eventually became fashionable. Occasionally even essential. And the long-decamped parasite living atop human heads took notice of this new layer of removable fur. A few head lice braved the new fibers for the chance to return to their ancient feeding grounds, where they have thrived ever since.

5

Who Shot the First Arrow?

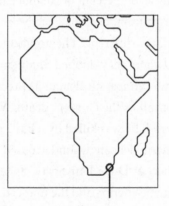

If the time of our species on earth
were a day, **this happened at 6:48 in
the evening** (64,000 years ago).

64,000 years ago

Invention of the bow

Some forty thousand years ago in the Zagros Mountains of
what's now northeastern Iraq, an elderly male Neanderthal was
hobbling on his arthritic feet near the Shanidar cave when an

unknown assailant struck a fatal blow. A small stone weapon pierced his chest, cleanly sliced through his ninth rib, and halted just short of his lung. When archaeologists uncovered his skeleton buried in the cave, they found healing on the bone that suggests he survived the initial strike, but the wound may have compromised his lungs or become infected, because the skeleton showed that within weeks he was dead.

By itself, there's nothing particularly unusual about a Neanderthal skeleton showing signs of trauma. They hunted the large mammals of the Ice Age, and a thrusting spear was their primary weapon. Inevitably, they suffered casualties. According to one study, as many as 80 percent of Neanderthal skeletons show evidence of severe trauma.

But this wound was different. The archaeologists who uncovered the skeleton (whom they dubbed Shanidar 3) noted that the groove in his rib was almost surgical in its precision, ruling out a mammoth tusk or fall as the cause of death. Most likely, the archaeologists believed, he was killed by a knife or a spear.

Still, the peculiarly delicate wound aroused the suspicion of a group of researchers at Duke University. Spears and knives deliver enormous kinetic energy, and the marks they leave on bones are typically jagged and comprehensive. The injury to this unfortunate Neanderthal was, by comparison, clean and precise. In order to test whether a stone knife or spear could have made such a wound, the team at Duke lined up a series of pig carcasses and stabbed, lanced, and flung spears at them with stone weapons. But no matter what angle they stabbed, or how softly they lobbed their spears, they couldn't replicate the trajectory or the precision of Shanidar 3's fatal wound. The blow, they concluded, was caused by a "low mass, low kinetic energy" weapon on a ballistic trajectory, suggesting the Neanderthal man wasn't speared, stabbed, or killed in a fall—but instead is the oldest victim of a bow and arrow archaeologists have ever discovered.

Although Shanidar 3 was a Neanderthal, he wasn't killed by one. There's no evidence the Neanderthals ever built or used any projectile weapons beyond spears—perhaps because they lacked the brainpower or perhaps because a bow may not have been useful against the big game they preyed upon. But whatever the case, there is only one species capable of inflicting a wound of this kind: our own, *H. sapiens.*

The weapon the murderer wielded would have been a multipart system, ingeniously designed and carefully built. It was a new kind of wooden catapult, strung with animal sinew to drive a light, perfectly straight arrow tipped with a stone the size of a thumbnail. Unlike a spear, which immobilizes its target by crushing muscle and bone, an arrow plunges into its victim to strike organs and arteries. While a spear retained its superiority against powerful animals, the bow remained the premier weapon against smaller targets—including humans—until the invention of gunpowder in the ninth century.

The bow is also far more complex than the spear. While the spear can be lopsided and still lethal, the bow must be perfect. The bow that shot Shanidar 3 would have been made of hardwood and carefully shaped. Any significant imperfections would have resulted in inconsistent flexion, rendering it either fragile or useless. The bowstring, likely made of dried sinew, would have had to hold under immense tension while remaining thin and smooth.

And as demanding as the engineering is for the bow, the arrow's might be even more difficult. An arrow released even from prehistoric bows is flung forward with more than fifty pounds of force and accelerates to speeds in excess of one hundred miles per hour. If the arrow is too stiff or flexible it won't oscillate correctly around the bow, and in order to fly accurately it must also be straight to a degree that seems unprehistoric. Any deviation greater than one sixteenth of an inch off center and the arrow will curve in flight.

The arrowhead must be small, so the arrow retains its balance, but chipped to a razor's edge, enabling it to slide through skin and bone. To attach the small stone arrowhead to the wooden shaft, the ancient archer would have carefully prepared a kind of prehistoric superglue made by melting and hardening tree sap.

The archer also needed to be well practiced. Small errors in angle correction, lead adjustment, or a poorly timed breath, and instead of a lethal tool delivered by stealth, the arrow headed for Shanidar 3 would have advertised the archer's position.

Whoever shot Shanidar 3 did so with a system of engineered perfection far more advanced than seems possible so long ago—which is an indication that Shanidar 3's killer wasn't the inventor. Instead, the original bow and arrow, like all new inventions, would have been rudimentary. It would have been light, inaccurate, and low powered. And, like his invention, the inventor himself was probably quite small.

I'll call him Archie.

Archie lived approximately sixty-four thousand years ago, maybe on the eastern side of South Africa in the Sibudu Cave just outside the modern city of Durban and where Marlize Lombard, professor of archaeology at the University of Johannesburg, recently discovered sixty-four-thousand-year-old stone arrowheads—the oldest evidence of the bow and arrow's use ever found.

Archie belonged to a cultural period archaeologists call the Howiesons Poort, and he lived among a group of hunter-gatherers along the South African coast who made some of the oldest pieces of symbolic art ever discovered: beads, shell necklaces, and small engravings.

Archie was a modern *H. sapiens.* If you dressed Archie in modern clothing, he would fit alongside other students in a contemporary classroom. He spoke a full language that we could learn to understand, just as he could learn to understand the words on this page. His skin was dark, and his hair wound in tight curls hewn closely to his head. He wore light animal skins, as much for decoration as for warmth, though he probably didn't wear shoes.

Archie would have eaten a mixed diet of tubers and meat— mostly small animals caught with small snares, but occasionally large ones such as buffalo. He also made art. There's an abundance of red ochre in the Sibudu Cave, which he would have crushed and mixed with water to make paint. No cave paintings have ever been found that date to this time, but there's little use for red ochre other than paint. If he didn't paint the walls, he may have decorated his body.

He bored holes in the shells and beads, presumably in order to string necklaces or perhaps to attach to sticks. These simple pieces of art, used as decoration or perhaps as toys, may answer one of the most perplexing questions in archaeology: How exactly did anyone come up with the bow?

The mystery of the bow's invention derives from its originality. Nearly every early hominin invention has parallels in nature, and the inventors, as ingenious as they were, mimicked what they saw. A rolling log inspired the wheel; a floating one inspired boats; from sticks came spears; from vines came ropes; from rocks came hand axes; from birds came planes. Nature often provides the idea, and humans provide the engineering.

That is decidedly not the case with the bow and arrow. Storing energy in the flexion of wood to launch a projectile does not replicate anything in nature. It is a uniquely human invention—which means it most likely began as did many of *H. sapiens'* most creative inventions: by accident.

Miriam Haidle is a professor of paleoanthropology at the University of Tübingen in Germany and a leading authority on the bows of the Howiesons Poort. "In my view," Haidle says, "it was invented by someone trying to do something else."

She's not alone in her belief. Many scholars have come to the conclusion that the bow is simply too novel, and too ingenious, for anyone to have come up with in a brainstorm or moment of inspiration.

When I asked Haidle how she thinks the bow may have been invented, she notes the existence of sticks, sinew, and beads in the Howiesons Poort and wonders whether someone like Archie, experimenting with tying beads to a stick, might have tied the two ends of the stick together and plausibly have arrived at a protobow. It would have been useless as a weapon, but the natural flexion would have been novel. Archie might have played with the tension in the system, drawing and snapping his creation to some delight. And if Archie used the snapping sinew to launch beads, shells, and sticks, the "bow" would have been state-of-the-art interesting, even if it yielded no practical value.

The path from an interesting but otherwise useless item to the sophisticated weapon system capable of killing Shanidar 3 would have required a long road of refinement and improvement from Archie's bead-flinging bow—from selecting and shaping the proper wood to sharpening and gluing the arrowheads to learning how best to cure the bowstring. Each of these small but essential innovations would have taken untold generations. So what explains the bow's persistence in the culture to allow for these improvements?

The reason may be simple: Archie was a child. A young boy, perhaps, because repeated studies suggest young male primates, from *H. sapiens* to vervet monkeys, prefer projectile toys. And there's reason to believe the first bows were novel and impractical objects of delight. In other words, the first bow was a toy.

Archaeologists are hesitant to use the word "toy" in reference to items found in ancient sites. As Patrick Roberts of the Max Planck Institute for the Science of Human History wrote to me, "We only really get firm evidence for toys back to urban periods in Egypt and Mesopotamia in 3000 BC"—in other words, after the invention of writing, when toys can specifically be described by their users as such. Before writing, it's difficult for archaeologists to differentiate dolls from idols.

Yet all children play. This is not only a human universal, it's a vertebrate universal. Stuart Brown, founder of the National Institute for Play, has devoted most of his career to studying the role of play in human and animal mental health. When I asked him if ancient children played, he told me that very few cultures have ever been studied that did not engage in playful activity, and "there's no reason to suspect that cultures before the advent of writing were any different, and in fact, there's some evidence to suggest they played more."

The popular conception of hunter-gatherers is one of peoples who scratched out a meager and brutal existence. Yet the best evidence suggests that hunter-gatherers, even those who live in arid environments, actually enjoy more leisure time than farmers and herders. Free time, play, and toys would have played as important a role in the Howeisons Poort period as they do to this day.

Not only can we assume a child like Archie would have played in a similar manner as exhibited by children all around the world today, but when I asked Scott Eberle, the former editor of the *American Journal of Play*, whether it's uncommon for weapons to have begun as toys, he told me the developmental path from toy to weapon is well traveled. Everything from boomerangs to robots to rockets began as toys, he said, before improvements in their manufacture made them effective on the battlefield. "My guess is that inspiration for designing weaponry travels close to the impulse to play," Eberle explained.

Modern child psychologists and the toy companies who hire them use the term "play patterns" when describing children's undirected activities. The list of activities isn't long, but near the top is one of the most fundamental: "launching and throwing," the enjoyment of which appears nearly universal. The young boys of the !Kung San play with miniature bows that fire miniature arrows. Meanwhile, as of this writing, six of the top twenty bestselling toys in America fire projectiles.

The first weaponized bows were far from the killing machines they eventually came to be. The Howiesons Poort culture likely used the bows found in the Sibudu Cave primarily for small animals, Haidle tells me, such as a kind of miniature antelope called the blue duiker, which weighs only ten pounds and whose bones litter their caves.

But unlike many early inventions, the bow provided a platform for near-continual improvement. *Homo sapiens* have added power and accuracy to the bow for more than sixty thousand years, turning it from a weapon that chipped Shanidar 3's rib to one that decimated the French at the Battle of Agincourt in 1415. It wasn't overtaken until another equally impractical curiosity—fireworks—proved to have lethal possibilities. But for nearly sixty-four thousand years, the bow was the dominant weapon of war.

And it all began as a toy.

6

Who Painted the World's First Masterpiece?

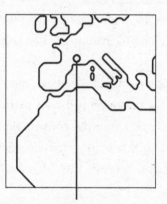

If the time of our species on earth were a day, **this happened just after nine at night** (33,000 years ago).

33,000 years ago

First masterpiece

In a cave located above the Ardèche River within sight of the magnificent Pont d'Arc land bridge in southwest France, brushed upon a limestone wall behind an entrance now secured by a modern steel door, is the oldest tangible proof of genius.

It's a painting. Archaeologists call it the Panel of the Horses, and it was created by a single painter whose work with a charcoal stick by lamplight thirty-three thousand years ago is now considered an example of unquestioned ancient genius.

The Chauvet cave contains more than four hundred paintings created over some five thousand years of periodic occupation until a landslide sealed off the cave's entrance nearly twenty-five thousand years ago. The cave of masterpieces remained hidden until Jean-Marie Chauvet and a group of cavers rediscovered it in 1994. Peter Robinson, an artist and editor of the Bradshaw Foundation, a group dedicated to studying and conserving rock art around the world, tells me the quality of all the cave's paintings is so consistently high it's likely only the best painters were allowed to paint its walls.

Yet in this cavern painted by masters, the Panel of the Horses stands above. It's a masterpiece by an artist who lived nearly thirty thousand years before the construction of the first pyramid, who worked quickly with a charcoal stick by the flickering of an ancient oil lamp to create the oldest example of prehistoric genius.

Who was the artist?

I'll call him Jean.

And because the only painter archaeologists have definitively identified in the cave stood six feet tall—judging by where he stamped his hand in the cave—I'll call Jean a him.

Jean was born nearly thirty-three thousand years ago in

modern-day France, which is nearly thirty thousand years before anyone ever wrote a word. Archaeologists call this era the Aurignacian, and the culture is characterized by its remarkably sophisticated art, including ivory beads, sculpture, and cave paintings, as well as some of the oldest musical instruments ever found.

Jean himself probably had dark-brown skin and curly dark hair with eyes that were most likely brown, according to DNA recently extracted from bones found in Belgium's Goyet caves.

He was completely modern in appearance. The prominent brow ridge on the faces of early *Homo sapiens* had receded by Jean's time, and there's no reason to suspect that he would stand out in any anatomical way if he stood on a street corner today.

But even if his face would be familiar, Jean was born into a Europe that modern eyes would hardly recognize. Ice sheets covered Scandinavia and the glaciers layered on the Alps were over a mile thick. Glaciers locked up so much seawater that Jean could have walked to England. To survive, he wore heavy bear and reindeer pelts that were sewn, fitted, layered, and lined with wolverine fur. Although European Ice Age hunters are known for mammoth hunts, the evidence suggests they rarely brought the large animals down. Instead, ibex bones comprise three quarters of the remains found in Aurignacian camps, suggesting the meat of the tough, agile animal similar to the mountain goat formed the foundation of Jean's diet.

Jean enjoyed music and might even have played a little himself. Archaeologists have found the oldest musical instruments ever discovered in Aurignacian sites, including small flutes made of ivory and vulture bone. Replicas of the flutes suggest they sounded like windy recorders and played five notes to the octave, which is a scale used in modern tunes such as "Baa Baa Black Sheep" and Led Zeppelin's "Stairway to Heaven."

Jean shared Western Europe with cave bears, rhinoceroses, reindeer, cave lions, and of course other groups of *H. sapiens,* though their interactions must have been sporadic on the almost unimaginably unpopulated continent. More *H. sapiens* live in Portland, Oregon, today than lived in all of Jean's Europe.

Jean, like all modern *H. sapiens,* was born able to draw. Studies performed on young children unexposed to drawing show they can both draw recognizable figures without any instruction and recognize drawings for what they represent. Drawing, in other words, is not an invention; it's an instinctual ability.

But for Jean's artistic talent to grow into true skill, it would have to have been recognized and nurtured from a young age. There's even reason to believe Jean was schooled. When I asked Robinson about the possibility of a formal system of education, he said Aurignacian paintings such as those in Chauvet are so well considered and so consistent in style there's reason to suspect something more than mere mimicry was at play, and that even in this ancient culture there may well have been an established system of apprenticeship. "Researchers talk about the possibility of a prehistoric 'school of art,'" Robinson wrote to me, and the evidence on the walls of these ancient caves suggests a far more formal system of education than you would otherwise suspect in such an ancient culture. Painting was not a frivolous pursuit, at least not when done in caves like Chauvet. "The art," said Robinson, "was taken very seriously."

This devotion in time and resources to art by a small hunting-and-gathering culture living in ice-encrusted Europe seems preposterous to those of us brought up on popular images of "cavemen" wearing tiger skins and spending their time either finding lunch or being lunch. But the best evidence suggests these images aren't just wrong, they may be totally backward. There's no reason to suspect Jean was any less likely to be a genius than a person born today, and there's at least some evidence to suspect the average person of his era was smarter. His brain, for example, was 10 percent larger than a modern person's, which could have been related to early *H. sapiens'* increased muscle mass over the average person today, or that our brains have become more efficient. But it can't be ruled out that he simply had more intellectual horsepower.

In order to survive, Jean needed a far more comprehensive understanding of his environment than we do today. He needed to track, hunt, build, butcher, cook, fight, communicate, socialize, and craft tools. He had to know which plant cured and which plant killed. He had to know where to find water, how to read the seasons by the paths of the stars, the migration patterns of the animals, their territory, and how to make everything he owned. And without writing, he had to commit it all to memory.

Let's speculate about Jean's age when he painted his panel. The word "genius" brings to mind masterworks produced at precocious ages, like Mozart composing his Symphony No. 1 at age eight. But the more common route is circuitous, according to a study by the Dutch economist Philip Hans Franses. Contrary to popular conceptions of genius, very few painters create their best work before their late twenties. Instead, master artists typically produce their most valued painting in their early forties. And while the UN's Population Division estimates the average life expectancy for a person of Jean's time was only twenty-four years, if a person in the Aurignacian managed to survive childhood it wasn't unusual for them to reach the age of fifty or sixty.

A reasonable guess would then be that by the time Jean painted his Panel of Horses, he had likely reached middle age. By then he would have not only mastered his art, but attained a measure of acclaim within his community, which granted him the rare honor of painting Chauvet.

When Jean first entered the cave, he would have undergone a singular sensory experience. On the light spectrum, a dark night is closer to broad daylight than it is to the complete lightlessness of a cave, which meant that once he moved beyond the opening, Jean would have been immersed in the rare condition of absolute black. To see, Jean used only pine torches or animal-fat lamps that produced no more light than a candle. As a result, he would never have seen the entirety of his tapestry. Instead, the cave art would have danced in shadow. Passing light would have revealed and animated the animals, so they may have appeared as if they emerged from the rock itself.

Sounds would have added to the mysterious aura. At times the cave would have been freakishly silent, and at others it would have gurgled to life as water trickled down roots and dripped from the stalactites in staccato slaps that echoed throughout the chambers.

Between the darkness, the flickering light, the shadows, and the sounds, it's easy to see how Jean could have viewed art as fundamentally alive and mysterious, and easy to imagine how caves could have combined church and theater to become places of supernatural power. They were not used as shelters. The absence of human bones, artifacts, or domestic remains makes that clear. Instead, Chauvet was a place to paint, visit, and quite possibly worship. Archaeologists will likely never obtain proof that religious ceremonies occurred in Chauvet, but a stone placed in the center of the largest cavern with a cave bear skull sitting atop it looks for all the world like an altar.

Jean would have begun his panel by first considering his canvas: crevassed limestone, a surface that offered both unique challenges and opportunities. The walls jutted and recessed, creating a lumpy, uneven surface, but one that a talented painter like Jean could use to create perspective and motion. He would have taken the wall's contours into account as he considered his subjects, which wouldn't be people or places. He would paint, as he always did, the great animals of his world.

One of the mysteries of Aurignacian cave art is why the artists almost never drew anything but animals. Out of the more than four hundred drawings in Chauvet, there's only a single, partially drawn human. There are no trees, mountains, or landscapes of any kind. Not even a hint of the massive, iconic Pont d'Arc land bridge nearly visible from the cave's entrance. It's not just that Aurignacian artists painted only animals; they only painted specific species. Jean didn't draw the animals he often saw, such as the ibex. Just as the squirrel and pigeon are neglected in our modern art and mythology, the ibex and rabbit are underrepresented in his. Instead, he drew the powerful and fearsome animals of his world: the rhinos, lions, cave bears, and bison. He drew the superheroes, not the food.

Jean prepared the limestone by scraping it of its calcite growth so his black charcoal would contrast against its white walls. He used charcoal of Scotch pine as his paintbrush, which he is somewhat more likely to have held in his left hand according to the Harvard neurologist Norman Geschwind's research on talented artists.

From the way the lines on the Panel of Horses intersect, Robinson believes Jean began by squatting low to draw the two rhinoceroses smashing their horns together at the bottom right of the tapestry. As he drew, finger marks on the wall suggest he periodically ran his hands over his work. The archaeologist Jean Clottes, author of *Cave Art* and among the first modern people to enter the cave, believes artists like Jean may have touched their paintings in order to connect with their spirits as they drew. In shamanism, communication between the supernatural travels in both directions, and Clottes believes Aurignacian artists may have drawn their animals partly to speak with their spirits.

Jean worked quickly. After his fighting rhinoceroses, he painted clockwise and to the left. First he drew a stag, then two mammoths, and then at the top, the aurochs—the wild predecessor to the domesticated cow. But in the middle of his work, in "a space he reserved for them," Robinson tells me, Jean drew his master-piece: his four horses.

He started at the top left and drew downward, each horse seemingly in full gallop from right to left, three with their mouths agape, panting, and the fourth in full whinny.

Finally, as a finishing touch, Jean engraved the outline of his horses, as if to highlight them in particular. Experts estimate by the technique, speed, and colors of their hides—produced with a mixture of charcoal and clay—the horses were all done at once, in quick succession, by the same artist.

When Pablo Picasso toured the cave art of the similarly spectacular Lascaux caves, discovered in 1940 in southwestern France, he is rumored to have said as he left, "They invented everything."

The list of artistic innovations created by Jean and other Aurignacian artists is long. Perspective, which uses size and angles to create the illusion of three dimensions on a two dimensional surface, was originally thought to have first been used by the Athenian artists and perfected in the Renaissance, but Jean employed it almost thirty thousand years earlier with his clashing rhinos and descending horses. Pointillism, the use of dots and space to create an image, is generally credited to French artists in the late 1800s, and yet near the entrance of the Chauvet cave an artist used pointillism to paint a red mammoth.

Stenciling and negative imaging are also common in Chauvet. The artists used thin, hollow bones to blow ochre onto the wall with their hand spread-fingered in place. The effect was to leave a shadow on the wall, a unique handprint that seems to reflect an ancient desire to say, "I was here." One handprint is so clear, a hand surgeon I showed the image to diagnosed his curiously crooked pinky as the likely result of a fracture of either the fourth or fifth metacarpal. "Healed," she said. "But imperfectly."

A few of the artists appear even to have employed animation. A bison on the walls of Chauvet has eight blurred legs, which is an idea that was hailed as new and innovative some thirty-two thousand years later.

If Jean spent his youth in a prehistoric apprenticeship, he might well have returned the favor as he grew older. Maybe he toured the cave with the next generation of painters. Near the back of Chauvet, leading into the altar chamber next to the carbon flakes of a torch, is an indentation in the mud where a young child stepped more than thirty thousand years ago and left one of the oldest footprints in the world.

As an elder man in the Aurignacian, Jean may well have died soon after he drew his horses. He could have fallen victim to an accident, infection, or violence. And because his culture likely believed in spirits and ritualized death, it's not implausible to hope a man so valued for his artistry was honored with art upon his death.

7

Who First Discovered the Americas?

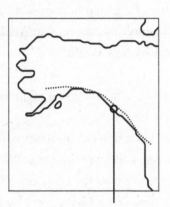

If the time of our species on earth
were a day, **this happened at 10:43
at night** (16,000 years ago).

16,000 years ago

First foot on the Americas

On October 12, 1492, when Christopher Columbus sighted what
came to be called the New World and claimed it for the kingdom
of Spain, an estimated 50 million people already lived there.

Clearly, his discovery merits an asterisk. Even calling Columbus the first European to discover the New World is problematic. Most researchers agree that the Norseman Leif Erikson landed on the Newfoundland coast nearly five hundred years before Columbus was born.

Christopher Columbus was more accurately the *last* person to discover the Americas. There was, however, a first.

Unlike the other inhabitable continents, the Americas are isolated from the African cradle of our species by ocean and ice. The result is that no hominin foot touched them until sixteen thousand years ago, when the Canadian ice sheets first began to melt and a member of a group of incredible adventurers living on the landmass connecting Siberia and Alaska placed the first foot in the unglaciated New World.

Who was it?

I'll call him Dersu, after the great eighteenth-century Siberian explorer Dersu Uzala.

Our Dersu was born sixteen thousand years ago, at the tail end of the last Ice Age and more than five thousand years before the agricultural revolution on the other side of the world, which means that at the time of his birth, every human on earth still hunted and gathered their food.

Dersu was born on Beringia, which is what archaeologists call the now-sunken, Texas-size piece of land that once connected Siberia and Alaska. Today it's sometimes called a land bridge, but it was hardly a Panamanian-size isthmus of land. Instead, Beringia is the closest to a sunken continent this planet can offer.

When Dersu was born, massive ice sheets had trapped enough seawater above land to drop ocean levels by nearly 300 feet, which exposed a vast piece of land that is now sunk 150 feet beneath the Bering Sea. When the early Spanish explorers theorized that the native people of the Americas reached the New World by walking across the mythical Atlantis, they were more right than they knew.

Little is known about Dersu's culture. The Bering Sea has drowned most of their archaeological sites, so what paleoanthropologists have pieced together is largely based upon language reconstruction, DNA from the earliest Americans, and the vanishingly few traces they left in Alaska and the northern Yukon. Yet it's indisputable that their perseverance is one of humankind's greatest survival stories.

Dersu was surrounded by uninhabitable environments. To his east lay the skyscraper-size Cordilleran ice sheet, which spanned nearly the entire western half of Canada. To his north and south the frigid Arctic and Bering seas, to his west the freezing deserts of Siberia. His home itself was one of the most formidable, inhospitable environments our species has ever settled. Beringia bordered the Arctic Circle during one of the coldest periods in human history. The average temperature was below zero, and in the winter the highs rarely rose above freezing. Shrubs were sparse and trees nonexistent, so archaeologists believe the Beringians relied on twigs and bone to fuel their fires.

Dersu was a fully modern *Homo sapiens* as smart as (and probably smarter than) you or me. According to archaeologist James

Chatters's reconstructions of the first Americans, Dersu may have more closely resembled the aboriginals of Australia than the Eskimo and Inuit.

Like any modern *H. sapiens*, Dersu would have danced, played music, and told stories. He spent time with his children, cooked his food, and filled his days hunting the kelp forest that ringed the northern Pacific. He ate seal, fish, shellfish, and, when he was lucky, the occasional wild horse. One of the few pieces of evidence the Beringians left behind is a horse jaw with a deep, stone-cut groove, which archaeologists found in a cave in the northern Yukon. The jawbone suggests that twenty-four thousand years ago in the Bluefish Caves a Beringian cut out and enjoyed a horse tongue.

Dersu wore furs he made from the skins of the seal, walrus, hare, elk, and moose that he hunted. His primary weaponry consisted of spears with sharpened flint and obsidian points and darts he launched with a type of spear-thrower called an atlatl. He fished with nets and ate the shellfish he gathered. He lived among the mammoths, but he probably didn't kill them as often as popular myths might suggest.

Dersu may have lived in a clan as small as just four or five families, an estimate supported by a twelve-thousand-year-old site in Sun River, Alaska. But he communicated, traded, hunted, and even paired off with his neighbors. Prior to the discovery of two young children buried together thirty-four thousand years ago near Moscow, most paleoanthropologists assumed isolated arctic hunter-gatherers were the product of at least mild incest. But DNA tests of the two children came to the surprising conclusion that they were only distantly related. It seems likely Dersu and the arctic tribes found incest as taboo as we do today and maintained relationships with distant tribes to avoid it.

Dersu must have occupied a position of some political authority within his group, a role typically filled by an elder man in arctic hunter-gathering cultures. Just by reaching middle age Dersu had already overcome long odds. Many ancient hunter-gatherer societies suffered an elevated rate of childhood mortality, but in the Arctic it may have been particularly bleak. Even in more technologically advanced arctic hunter-gatherer groups, as many as 40 percent of children died. In Beringia it may have been a coin flip whether a child reached their tenth birthday. Dersu may have had the good fortune to survive into adulthood, but if he fathered children he would have likely suffered loss.

At the Sun River site, archaeologists found three buried bodies. They're all children. A few families used the seasonal homes for years, the archaeologists believe, but after they cremated a three-year-old in the center of their hut, they abandoned the site for good. A hint, perhaps, at prehistoric heartbreak.

Dersu is one of the first Beringians born for whom traveling south was an option. For fifteen thousand years—a period three times longer than all of written history—the half-mile-thick Cordilleran ice sheet blocked any movement south. Until recently, most scholars believed the founding New World population arrived through central Canada some thirteen thousand years ago, when a pathway melted between the Cordilleran and Laurentide ice sheets along their meeting point at the Continental Divide. But recent discoveries of an earlier American presence have all but laid that theory to rest. In the last decade, archaeologists have turned up significant evidence of human occupation in Oregon and Chile dated before the Great Divide path opened. Many archaeologists, including University of Oregon archaeology professor Jon Erlandson, are now convinced humans arrived at least sixteen thousand years ago. At that time, the only viable route would have been a treacherous journey in small boats down the western Canadian coastline. Recent surveys of Canadian coastal boulders have confirmed a few first peeked out from the ice sheet some sixteen thousand years ago, suggesting a pathway existed.

When I spoke to Erlandson about the possibility of a coastal journey, he told me he believes the thriving kelp forest that ringed the western edge of the Canadian coast not only fed the Beringians, but laid a southern path for Dersu to follow. Erlandson's research has led him to believe this "kelp highway" ran from Beringia all the way to Baja, providing both food and incentive for Dersu's journey.

Yet Dersu would have lacked the large, open-ocean boats required to make the entire trip in a single voyage. The giant trees needed for such a vessel simply didn't exist in Beringia. Instead, his boats probably looked like modern-day kayaks with animal skins stretched across a driftwood or whalebone frame. These kayaks couldn't have supported multiday journeys, but Dersu could have used them to fish, hunt, and paddle around the fingers of ice that still protruded into the sea.

Even if archaeologists have largely pieced together how Dersu and the first Americans made their journey, the question of why remains elusive. To anyone living today in the comfortable climates south of the 40th parallel, it might seem obvious. But Dersu did not have the benefit of a map. He couldn't have known if what lay ahead was an impassable stretch of ice or sea, or if where he headed was far worse than where he left. There's no evidence the Beringians had to leave—either at the tip of a spear or at the brink of starvation—as arctic hunter-gatherers continued to thrive in Beringia and Alaska for thousands of years after Dersu.

Without any obvious imperative forcing Dersu's departure, the best explanation for his voyage may be no more complicated than his desire to explore. Exploration might seem like the frivolous pursuit of a modern person, but the evidence suggests this desire is as old as, if not older than, *H. sapiens.* Dersu and his people may have left simply because they wanted to see what lay beyond the horizon.

The desire to explore is an offshoot of the evolutionary requirement to see, find, feel, model, and expand the boundaries of our world. It's baked into the human condition. There's no reason to believe Magellans or Neil Armstrongs were any less common in ancient times than they are today. If anything, the rewards of untouched islands and virgin continents may have compelled even more exploration. When archaeologists uncover evidence of dangerous journeys like Dersu's, there's a tendency to suspect an enemy or starvation forced the voyagers to take such a dangerous risk. But careful reconstruction of other great seafaring cultures has suggested this isn't necessarily the case.

Archaeologists reexamining the movements of an ancient South Pacific culture of navigators called the Lapita found they discovered and moved to multiple islands within single generations. They colonized islands so rapidly that archaeologists now believe the best explanation for their relentless yet incredibly dangerous search was the basic human desire to explore the unknown.

Archaeologists don't know how quickly Dersu made the journey down the coast. It could have been within one generation, or it could have been over the course of many. It may have been Dersu's children or grandchildren who first breached the southern glacial barrier. But once beyond the ice sheet, Dersu or his descendants would have discovered an untouched Eden of megafauna including lions, camels, mammoths, mastodons, cheetahs, horses, 125-pound beavers, and giant condors. In all, when Beringians first arrived, ninety separate species weighing more than one hundred pounds walked, crawled, or flew in North America. But unlike the large animals of Africa that coevolved with hominins, the New World megafauna were unadapted, unafraid, and unprepared for the new superpredator. Within four thousand years, humans hunted all but a few of them to extinction.

The new Americans spread so quickly and thoroughly across the Americas that some linguists believe nearly every Native American language spoken south of Canada originated from a single mother tongue. Despite what flowered into remarkable linguistic diversity, the linguists Joseph Greenberg and Merritt Ruhlen have found remarkable similarities in the diaspora of languages across the Americas, and based on these similarities have reconstructed portions of what they believe may have been their mother tongue: the language, in other words, that Dersu spoke. If Greenberg and Ruhlen are correct—and they certainly have their detractors, though I find their methodology compelling—then because of the unique way the Americas were populated, Dersu could be the first person whose words we know. So when Dersu, who navigated treacherous arctic seas in the shadow of glaciers to discover the new world, passed away, we may know one of the words he said.

His word for "death," Ruhlen believes, sounded like "MA-ki."

8

Who Drank the First Beer?

If the time of our species on earth were a day, **this happened at 10:48 at night** (15,000 years ago).

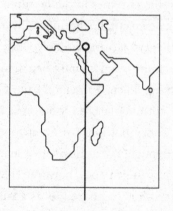

15,000 years ago

First beer

On a September morning in 1795, mutiny broke out aboard the Royal Navy's HMS *Defiance*. The 74-gun ship of the line had recently returned from a long and chilly voyage through the

Baltic Sea, but despite making port near the Scottish city of Edinburgh, Captain Sir George Home continued to serve his men a watery grog that was "as thin as muslin and quite unfit to keep out the cold," according to one disgruntled sailor's report. After months of drinking the unsatisfying brew the crew revolted and stormed the captain's quarters.

The insurrection took two days and the assistance of another ship to finally be put down. Five sailors were eventually hanged. The mutiny over the thin grog was far from the first time alcohol drove hominins to make an irrational decision. According to geneticists, our dogged pursuit of alcohol began long, long ago.

Studies suggest the enzymes in the hominin gut enabling our ancestors to break down ethanol dramatically improved approximately 10 million years ago, presumably when the common ancestor of gorillas, chimps, and humans began to spend more time on the ground eating fallen, fermented fruit. Geneticists believe natural selection not only favored those apes able to digest the ethanol in these boozy but otherwise nutritious fruits, but also favored those who sought out its distinctive smell and taste.

These were not drunken apes, however. The alcohol content in fermented fruit is low compared to its volume, so this was a food, not a drug. The intoxicating effects came much later, when hominins learned to concentrate the juice of fermented fruit to make wine and to add water to honey to make mead. The formula for both wine and mead is so simple they may well have been made before *Homo sapiens* evolved.

Yet neither wine nor mead had the dramatic effect on society that beer did. Beer is made from cereals, which means it can be stored, predictably located, and harvested in mass quantities. With the discovery of beer, alcohol—for the first time—became available on demand. Given what we know about alcohol's motivating effect, it should be no surprise this discovery represented a pivotal moment in human history. Yet the discovery of beer takes on even more import because cereals are not only a source of alcohol, they're also a food. Today they provide nearly half of the world's calories, and their intensive cultivation sparked the agricultural revolution—the first transition from hunting and gathering to farming and herding—which to this day remains the most consequential shift in human history.

Because the first farmers worked harder and lived shorter, unhealthier lives than hunter-gatherers, many scholars have long believed that no one chose to farm. Instead, they believe hunter-gatherers living in Mesopotamia were lured into it like a lobster entering a trap. The bait, however, may not have been the pursuit of bread as many long believed. Instead, a growing body of evidence suggests it was beer, which would make the first brewer one of most important people in human history.

Who was it?

I'll call her Osiris, and a her, I'll propose, because she brewed her beer from seeds of either wheat, barley, or rye, all of which would have been more likely gathered by a woman in her hunter-gatherer community.

Osiris was born nearly fifteen thousand years ago in a small village in the Middle East, likely in a site like the one archaeologists call Shubayqa, in northeast Jordan, where in 2018 a team of archaeologists discovered the oldest baked cereals ever found.

Osiris was an early member of a group archaeologists call the Natufians, who were among the first people to live in the same place throughout the year. Yet she was not a farmer. Until the 1970s, most archaeologists believed farmers were the first to settle into year-round residences. But discoveries of multiple sites in and around the Euphrates Valley have proven the opposite: hunter-gatherers like Osiris lived in one place for thousands of years before they began to farm.

As a Natufian, Osiris likely stood roughly five feet tall and was darker skinned with brown eyes and black hair, according to geneticists' estimates. She made her home in a circular, semiunderground stone building with a stone foundation and wood walls in a village with a small handful of similar buildings. The entire population of Shubayqa was less than two hundred, yet because it was occupied year-round it was one of the largest cities in the world.

She would have decorated herself with small stone, shell, and bone ornaments. She used ostrich shells as containers, fashioned fishhooks and harpoons out of bone, made animal and human figurines out of limestone, and took part in feasts to honor her dead. Archaeologists have discovered the grave of one Natufian woman—thought to have been a kind of shaman—buried with eighty-six tortoiseshells.

Osiris lived in a time and place of bounty. Her home was far less arid than it is today, and yearly runoff from the nearby Druze mountains turned the plains into a semipermanent wetland. Herds of auroch and gazelle gathered nearby to drink from the water, while the fertile highlands supported legumes, almonds, and pistachios. Archaeologists still debate exactly why the Natufians settled into permanent villages, but the simplest explanation is that they lived in an area of year-round plenty.

One inevitable consequence of their sedentary life was their gradual accumulation of heavy items that would have been impractical when life was spent on the move. These heavy tools included harvesting equipment and kitchen appliances like stone grinders, sickles, and forty-gallon limestone cauldrons. They opened up new possibilities and new food sources, principally the gathering and grinding of seeds—and eventually, the brewing of beer.

Osiris spent her days gathering the plentiful fruits, nuts, and tubers near her village. Occasionally she would have harvested the wild ancestor of wheat, but it probably wasn't a staple because ripened wild wheat bursts its seeds onto the ground, forcing her to stoop for each one. Eventually, as humans repeatedly selected for the tough-rachis gene mutation in wheat plants that kept ripened seeds within their shell, we changed the nature of the plant. But in the time of Osiris that development had yet to occur. As a result, the caloric returns for gathering wild wheat would have been, in the words of paleobotanist Jonathan Sauer, "pitifully small." Before the discovery of beer, hunter-gatherers rarely bothered with it.

When Osiris gathered her wheat seeds, she was probably looking for something else—and happened to stumble upon a wheat plant with the rare tough-rachis mutation. Without having to gather the ripened seeds off the ground, she could have threshed a bushel directly from the plant, as is done modernly.

Using the seeds, Osiris would have made herself the Natufian version of gruel by pounding them from their shell and then soaking them in water to convert the cereal's starches to sugars. Maybe, if she had been lucky in her gathering, she would have added some honey or fruits to sweeten it. Once she made her bowl of gruel, it would have been a short step to beer. All that was required was a moment of forgetfulness, a stray fleck of yeast, and the hot Middle Eastern sun.

By one definition, beer is simply rotten gruel. The recipe to change gruel into beer is simple: All that's required is time, heat, and a yeast like *Saccharomyces paradoxus* or *cerevisiae*, which converts the sugars of the grain into alcohol and carbon dioxide. Fortunately for Osiris, the required yeasts are everywhere. They are in honey, so if she used some to sweeten her gruel, that would have kick-started the conversion. They are on acorns, so if Osiris had used the same pounding rocks for both her acorns and wheat, *that* could have begun the process. They're on insects, so if the right one landed on her gruel as it sat, that could also have started the fermentation. They even exist on the wild cereals themselves, Martin Zarnkow, professor of food and beer technology at the University of Munich, tells me. So if she were willing to accept a "bad yield," he says, Osiris wouldn't have needed to add anything at all.

Whichever the case, if she forgot her gruel in the summer heat it would have taken as little as a day to turn into a deliciously fermented mistake.

What would Osiris have tasted?

Scott Ungermann, brewmaster at San Francisco's Anchor

Brewing Company, tells me the rotten gruel—or light beer, as it would be called today—would have had a sour, acidic taste from contamination by a bacteria called lactobacillus, which enters unsealed brews and produces a lactic acid by-product. Modern breweries usually go to great lengths to avoid this kind of contamination, but in the case of a few sour brews, they introduce it intentionally, which means a few modern beers might even approximate the taste of the original.

The closest beer to Osiris's drink still brewed today, according to Ungermann, is called a Berliner Weisse, which is a light, sour beer made without hops. Berliner Weisses are "refreshing oddities," according to *Beer & Brewing Magazine*, "slightly hazy and effervescent, with a light tang." However, the Berliner Weisse lacks the solid wheat that would have floated atop Osiris's brew, which is probably why Mesopotamian pictographs suggest they drank their beer through a straw.

Osiris's beer would have had only half the alcohol concentration of even a modern light brew, so she wouldn't have become drunk. But she probably would have recognized and enjoyed its buzzy effect.

So she would have made it again. And shared it. And even though gathering undomesticated wheat, barley, and rye was incredibly inefficient, beer—unlike bread—was worth it. If not for herself, then for the villagers next door.

Previously, when they lived nomadically, the pre-Natufians could simply move away from a troublesome neighbor. But for Osiris, who had permanent buildings and stockpiles of food to protect, this was no longer an option. As a consequence, archaeologists have found evidence that the Natufians frequently threw their version of the modern-day block party, which they held both to ease tensions and to build social bonds. Under these circumstances alcohol, the great social lubricant, would have become all the more essential.

As Osiris and the other settlers of the Levant gathered wild cereals for their beer, paleobotanists speculate seeds would occasionally fall to the ground as they walked home. Over the generations, fields of wheat, barley, and rye sprouted near their villages. As these fields grew into a viable food source, the Natufians began to care for them. They weeded them, tilled their soil, and replanted the seeds. Eventually, what had begun as an inefficient food source grew into fields of productive, dense calories. At first people probably cared for these fields some of the time. Then most of the time. Then full time.

Because fields of crops can sustain far larger populations per square foot than hunting and gathering, the Natufian populations grew beyond what their old lifestyle could support. After a few generations of farming, the Natufians could no longer return to their previous ways without risking starvation. The pursuit of beer—and eventually bread—may have locked the Natufians and their descendants into an entirely different way of life.

The theory that beer rather than bread could have been the driving motivation behind the agricultural revolution is not new. It was first proposed in the 1950s by archaeologists like Robert Braidwood at the University of Chicago, but at the time the theory was largely dismissed. Paul Mangelsdorf, a Harvard botanist, summed up the general consensus in his response to Braidwood's 1953 article, "Did Man Once Live by Beer Alone?" when he wrote, "Are we to believe that the foundations of Western Civilization were laid by an ill-fed people living in a perpetual state of partial intoxication?"

But in 1972 the Nobel Prize–winning geneticist George Beadle solved the ancient mystery of corn's wild ancestor, and his discovery inadvertently tipped the balance of the beer versus bread debate. Beadle's discovery proved *H. sapiens* were capable of domesticating a vital food source in the quest for booze by showing humans had done it again. This time on the other side of the world.

Corn now provides the third-most calories of any crop in the world, but unlike wheat, domesticated corn bears almost no resemblance to any plant found in the wild. Artificial selection has so radically changed corn that its wild ancestor remained unclear until Beadle's genetic tests confirmed modern-day corn originated as the wild Mexican grass teosinte. This discovery forced anthropologists to reconsider why the Natufians first gathered cereals, because while wild wheat could have plausibly been gathered for bread, no one, no matter how hungry, would ever have bothered to gather teosinte.

Teosinte is a useless food source. Like corn it has ears, but you wouldn't know it without a microscope. An entire ear of teosinte contains fewer calories than a single kernel of modern corn. University of Michigan archaeologist Kent Flannery called teosinte "starvation food," and the University of Wisconsin botanist

Hugh Iltis called the protective case on wild teosinte "so inde-structible that human use of the grain is out of the question." He goes on to wonder why "anyone would bother to collect or try to grow this utterly useless grain when the grain itself is so perma-nently imprisoned."

The answer is alcohol.

The husk of teosinte is sweet, so it's thought that people living in what is now Mexico may have occasionally chewed the husks as candy. Eventually, some seven thousand years ago, they began to gather and press the cobs into a sweet juice. From there, teo-sinte juice needs only time to ferment into corncob wine.

As corncob wine makers artificially selected for larger teo-sinte ears and scattered their seeds, the cobs naturally grew. But it took at least three thousand years of artificial selection before teosinte ears were even two inches long, which is the first time it could plausibly have been used as food.

Because domesticated wheat, barley, and rye are genetically similar to their wild ancestors, natural selection worked far more quickly to turn them into productive foods. But the evidence, and human nature, suggests the process began the same way: with someone's discovery that a plant could produce alcohol.

Who drank the first beer? The ancient Egyptians, who brewed at least seventeen varieties and even branded them with names like the "heavenly" and the "joy bringer," had an answer. They believed beer was a gift from Osiris, their benevolent god of the afterlife. According to their belief, Osiris made a meal of water and sprouted grain and forgot it in the sun. The next day he discovered it had fermented but decided to drink it anyway. He was so pleased with the sensation, the story goes, that he passed it on as a gift to humankind. Which, around ten thousand years before the first Egyptian hieroglyph, is almost exactly how it occurred.

Only Osiris wasn't the god of the afterlife. She was probably just a young woman living in the Middle East who left out her lunch.

9

Who Performed the First Surgery?

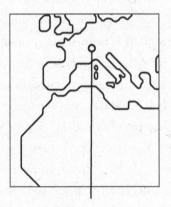

If the time of our species on earth
were a day, **this happened 33 minutes
before midnight** (7,000 years ago).

7,000 years ago

First surgery

In 1865, in the Peruvian town of Cuzco, a wealthy collector by
the name of Ana María Centeno showed American diplomat
E. George Squier a curious Incan skull she had come upon. The

skull was perfectly normal in every way save for a large rectangular hole cut so precisely above the ear that no weapon of war or animal's tooth could have been responsible. It was as if someone had carved a window to peer into the brain.

Centeno's skull wasn't the first time archaeologists had discovered a head so curiously carved, but until then the cuts had been dismissed as gruesome wounds of war or postmortem modifications. Squier had a different, far more radical theory: He proposed the skulls weren't cut to kill the owners—they were made in an effort to save them.

The self-taught Squier was, in a peculiar way, the perfect person to interpret Centeno's find. Squier was a fervent American patriot in both his diplomacy and his archaeology. When Abraham Lincoln appointed him to settle a financial dispute with Peru in 1863, he used the assignment to further prove his belief in the sophistication of the ancient cultures of the New World. This disposition helped Squier see Centeno's skull for what it was: an ancient example of medical genius.

Squier presented Centeno's skull to the respected French surgeon and anthropologist Paul Broca, whose conclusions shocked even Squier. Broca declared that not only had an ancient surgeon performed the procedure during the life of this person, bone growth on the wound's edges proved the patient had survived it.

The procedure is called a trepanation, but it is not brain surgery. In a trepanation, the surgeon removes a section of skull, but doesn't penetrate the membrane encasing the brain. If ancient surgeons with unsanitized tools had touched gray matter, the patient would have quickly died of infection. Broca's discovery of the patient's survival proved the ancient surgeon cut through the bone and halted.

This pronouncement shocked both the medical and archaeological world. At the time, two thirds of patients undergoing the modern equivalent of the procedure in European hospitals died. Yet this ancient man, whose surgeon had used a stone blade, lived. And he wasn't the only ancient surgical success. Archaeologists began to reclassify older skulls found in Europe and Russia as patients of trepanations, and in September 1996, in the city of Ensisheim in eastern France, archaeologists uncovered "Burial number 44"—the well-preserved skeleton of a fifty-year-old man who received not just one trepanation, but two. Grave goods buried beside him revealed he died more than seven thousand years ago, making this elderly man's skull the oldest discovered evidence of surgery of any kind ever performed. His head potentially represents the site of *Homo sapiens*' "first cut," the surgical term for a new procedure's first incision.

Who performed it?

I'll call him Dr. Zero, and a him because there's some evidence surgery was the direct result of a new social hierarchy in Neolithic Europe and may have been performed by male political authority figures.

Dr. Zero was born approximately seven thousand years ago as a member of what scholars call the Linear Pottery culture. He lived as a farmer and herder in a settled village in what is now eastern France, along the Rhine River only a few miles from modern-day Germany. He was one of the first farmers in Western

Europe, but neither he nor his ancestors were former European hunter-gatherers. Studies of ancient DNA have largely determined that as agriculturalists moved into Europe, they pushed out or exterminated the local hunter-gatherers rather than assimilating them.

His community lay at the frontier of this agriculturalist push. His people grew wheat, peas, and lentils; herded cattle; and occasionally hunted deer. The farmers and herders used the fertile ground along the floodplains and former pathway of the Rhine to support their crops. The village itself consisted of a handful of long houses, approximately one hundred feet in length and made of massive oak posts with slanted, thatched roofs and rows of poles in support.

Dr. Zero was short—men of his time and place averaged five foot four—as a result of his unbalanced, cereal-intensive diet that would have been as hard on his growth as his teeth. He would have been among the first fair-skinned Europeans, and DNA suggests he most likely had brown eyes with dark hair and was lactose intolerant. His dependence on a few food sources left him vulnerable to flood, drought, and disease. Starvation was an ever-present risk, and his health was far worse than his hunter-gathering neighbors, who benefited from a varied diet.

He made his tools entirely of stone, wood, sinew, and other organic material. He chopped trees with adzes—sharpened stone ax-heads affixed to wooden handles—and used sharpened flint and obsidian rocks as knives and arrowheads. Today, archaeologists describe his era as Neolithic Europe, the age of the continent's first farmers.

Dr. Zero must have been an authority figure within his community. As Stanford professor of archaeology John Rick tells me, "To convince someone that carving through their skull is a good idea, one must have some measure of authority over them."

When I asked Rick why there's no evidence of surgery before Dr. Zero, he told me the entire concept of authority—by way of anything other than sheer physical size—is a rather recent development in the history of *H. sapiens*. Our modern faith in someone's expertise derives from what anthropologists call specialization. Today, trust in the authority of others forms the basis of most professional relationships, but it's a relatively new concept that didn't begin until farming allowed one person to produce food for more than themselves. This food surplus enabled some members of society to specialize in different fields, such as military, governance, or medicine. The more efficient farmers became, the more people were freed to pursue increasingly specific trades.

The transition to farming and herding also marked the beginning of wealth. For the first time in human history, one person could have more than someone else. They could have more cattle, for example, or more crops. Thirty percent of the graves in Ensisheim contain elaborate seashell headbands, necklaces, and tools, while the rest have nothing. Archaeologists believe this is evidence that some people in Ensisheim were more respected than others, if not outright wealthier. Farming led to income inequality, and if someone has more than you, the inevitable tendency is to believe they know more than you, too, Rick tells me. The wealth gap combined with increasingly specialized professions, according to Rick, introduced and accelerated the concept of authority.

It was this introduction of authority, rather than tools or intelligence, that led to the emergence of surgery. The absence of supreme authority prior to farming may explain why archaeologists have yet to discover concrete evidence of surgery prior to the agricultural revolution.

But even if Dr. Zero had the authority to perform surgery, why would he want to? For years after the discovery of trepanations, archaeologists believed Dr. Zero and the rest of the ancient surgeons were simply quacks and curious butchers who belonged to the long and sordid history of premodern surgeons who disguised their depravity or ignorance behind an apron and scalpel.

But the sheer number of trepanations found across the world complicates what would otherwise be a straightforward explanation. Archaeologists have found trepanned skulls in Europe, Russia, Oceania, and South America spread out over nearly seven thousand years. If it was only Dr. Zero who performed the surgery, or if it was only performed in a specific time or place, it would be easy to assign his motivation as being based upon a particular religious belief or as another method of radical body deformation, like foot-binding or neck elongation. But ancient trepanations occurred in cultures that had no contact with each other and precious little in common, which rules out local customs or style and requires a more comprehensive explanation. As the medical historian Plinio Prioreschi wrote, "It was an activity rooted in experiences and needs that were common to all prehistoric men in all locations."

Dr. Zero's motives when he made his first cut require a more universal explanation than a local style or religion. And the archaeologist John Verano, author of *Holes in the Head*, has a particularly radical one: He believes these ancient surgeons practiced good medicine.

"Good medicine," if it's defined as effective treatment, is thought of as a rather recent development. As the Harvard biochemist and medical historian L. J. Henderson declared in the early 1900s: "Sometime between 1910 and 1912 in this country, a random patient, with a random disease, consulting a doctor chosen at random, had, for the first time in the history of mankind, a better than fifty-fifty chance of profiting from the encounter."

The history of surgery is even bleaker.

Dr. William Keen, a surgeon in the Union Army, wrote that by his calculation treatment in a city hospital was seven times more dangerous than fighting in Gettysburg. The year Broca declared the trepanned person had survived, doctors in London killed 70 percent of their patients who underwent a similar procedure. The idea that Dr. Zero performed good medicine and good surgery would be a radical departure from most of written medical history. But the procedure he performed does save lives today. It's called a craniotomy, and surgeons perform it when a significant head injury results in bleeding within the skull. As the brain swells, the pressure inside the braincase builds until it's eventually choked of its oxygen. The only way to relieve the pressure is to remove a portion of the skull.

According to Verano, more than half of the trepanned skulls

found in Peru, where archaeologists have discovered the greatest number, show evidence of earlier fracture. And that's likely an underestimate, as trepanation often removes the fractured bone. The patients were also predominantly male, and the trepanations more frequently performed on the left side of the head—indications that most patients suffered their injuries in battle, after blows from right-handed opponents. The evidence is strong, if circumstantial, that Incan surgeons primarily performed their trepanations in response to traumatic head injury.

Dr. Zero would have been familiar with head wounds, particularly those delivered by adzes. Mass graves were disturbingly common in early European Neolithic communities, as was blunt force trauma to the skull. A seven-thousand-year-old grave in Talheim, Germany, is filled with the bodies of thirty-four men, women, and children, fourteen of whom were killed by a blow to the head. The Schletz-Asparn grave in Austria has more than three hundred bodies, and another in Herxheim, Germany, has more than five hundred. Early Neolithic cemeteries in Ile Tévic, France; Vedbaek, Denmark; and Skateholm I, in Sweden, that allow for a comprehensive accounting of cause of death, suggest as many as 15 percent of Neolithic Europeans died violently. The evidence indicates ancient warfare and slaughter were common, suggesting that Dr. Zero would have been familiar with gruesome skull wounds.

The initial inspiration for trepanations may have come from cleaning cracked skulls of bone, scalp, and blood. As Dr. Zero became familiar with picking out shattered pieces of skull, he may have first observed damage to the gray matter was invariably fatal.

Secondly, he may have noticed victims with dented skulls suffered a series of worsening symptoms, including vomiting, confusion, an inability to speak, partial paralysis, and eventually death, while those with open wounds recovered more frequently.

Eventually, Dr. Zero must have arrived upon the remarkable connection that in the case of massive head injuries, a hole in the skull actually improved a patient's chances—and the next time he was presented with these horrific symptoms, he made the radical decision to cut the hole himself. It's exactly what a modern hospital would do today.

The moment a surgeon draws their scalpel across the skin of a patient is dramatic even in a modern operating room. Once the surgeon pushes their scalpel through a patient's skin they have committed themselves to the procedure, and they cut with the faith that the grievous harm they inflict will, in the end, be better than the alternative.

Dr. Zero would have made humankind's first cut with a chipped flint or obsidian blade, both of which can be as sharp as any modern surgical tool. He first would have removed a section of scalp, which would have produced a prodigious amount of blood and some pain, but can be done quickly. Once past the scalp he would have gone to work on the bone, but the pain for the patient would ease as there are no nerves in the skull. One reason trepanation could have been a common ancient surgery is the relative ease for both the surgeon and the patient. Surgery upon organs or soft tissue would have resulted in tremendous bleeding and serious risks of infection, while a trepanation is a relatively straightforward procedure.

Verano notes that there were a few methods used by ancient surgeons to cut through the skull, but the most successful one, judging by ancient survival rates, is the "shaving method." Rather than drilling through bone, which risks the surgeon punching through into the dura mater, Dr. Zero may have whittled away at his patient's skull with small cuts.

Once past the skull, Dr. Zero would have been presented with a semihardened Jell-O-like pool of coagulated blood—called a subdural hematoma—pressing into the dura mater. After Zero removed it and relieved the pressure, the results would have been dramatic. The patient's neurons on the surface of their brain would fire again, the confusion, paralysis, and slurring would cease, and any witnesses would think they had witnessed a miracle. Unlike in modern craniotomies, the skull couldn't be replaced with a titanium plate, so the patient would leave the operation with a depression in their head. But the scalp would eventually heal over and some patients lived for years afterward.

In the long run, trepanation's success might even have become its curse, because it's clear the method quickly became overprescribed. A few skulls suggest surgeons used trepanations to treat types of head pain they would have been ill-suited for. In the case of one trepanned child in Peru, archaeologists can see inflammation in the skull indicative of a middle-ear infection. The severe pain the child experienced seems to have inspired the surgeon to trepan, because the surgeon cut the skull in the area of the infection. Unfortunately, not only would trepanation have been ineffective, it would have made her condition considerably worse. The hole would have allowed the infection to spread onto the dura mater, probably resulting in a lethal case of bacterial meningitis. A lack of bone healing on the skull shows the child did not survive.

But even when the procedure was warranted, the long-term postoperative outlook might not have been entirely positive for many of Dr. Zero's patients. Devastating skull fractures that require trepanations can have lasting impacts on the survivors. According to a study published in the journal *Neurology* in 1985, over half of the victims of penetrating skull injuries developed epilepsy. An onset of seizures could explain why Zero trepanned his patient a second time, but unfortunately trepanation is ineffective for epilepsy, and the patient would eventually have died.

When Dr. Zero's time came, the members of his community may well have marked his passing with a ceremony or grave gift. Those typically included headbands, necklaces, and tools, but perhaps the citizens of Ensisheim left Dr. Zero with something different, a piece of sharpened obsidian: history's first scalpel.

10

Who First Rode the Horse?

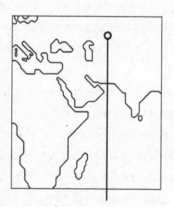

If the time of our species on earth were a day, **this happened 27 minutes before midnight** (5,600 years ago).

5,600 years ago

First horse ride

In 2006, at an archaeological site in northern Kazakhstan, home of an ancient culture known as the Botai, the zooarchaeologist Sandra Olsen and her team dug through ancient human trash pits testing the soil until they struck the horse historian's version of a goldmine: a thick layer of 5,600-year-old horse manure.

Because no culture, past or present, has ever bothered to throw away the feces of wild animals, Olsen's discovery is the first evidence of the horse's domestication. A Botai garbage pit containing horse manure is the equivalent of a modern-day trash can filled with dog and cat poop but no bear scat. If a future historian dug it up, they would conclude that we lived with dogs and cats, but not bears.

The ancient horse manure in the Botai camp is the oldest proof humans had domesticated the horse, but many researchers suspect the Botai acquired their horses from older herding cultures to their west who had a history of animal domestication. Exactly who first domesticated the horse is an unsettled debate among scholars.

What is settled is the way they were first used. It wasn't for riding, but for food. Horses were a source of meat and milk, and cultures of the high steppe prized them for their ability to graze through snow, which their sheep and cattle could not. There's no evidence these first domesticators ever rode their horses, probably because they had no means of control. They had no saddle, no stirrups, and most important, no bridle. Without any way to turn or stop, riding a horse would have been a short, potentially bone-breaking experience. It may have been done occasionally as a stunt, but horses without bridles would have resembled motorcycles without handlebars: exciting, but poor transportation.

The Botai changed that.

In 2009, working in a Botai archaeological site, the anthropologist David Anthony found a set of curiously worn horse teeth that sit alongside Neil Armstrong's boot print on the moon and the first wheel-rutted road in Germany as physical proof of the most significant moments in the history of human transportation.

The simple abrasion—amounting to a few shavings taken off the horse's front molar—is, as Anthony later determined, the exact and distinctive damage that occurs when a horse manipulates its bridle.

Worn horse teeth sound anticlimactic compared to prints in moon dust or ancient roads, but these ground teeth prove that for the first time in human history, someone traveled over land faster than their legs could carry them.

Whoever invented the bridle didn't just improve land transportation—they invented it.

Who first rode the horse?

I'll call him Napoleon in honor of Napoleon Cybulski, the Polish physiologist who first isolated adrenaline, a molecule that played no small role in this moment of inspiration.

Napoleon was born in northern Kazakhstan nearly six thousand years ago, which is about the time when, thousands of miles to the west, bronze was first forged, early cities formed, and scribes were even scratching some of the first words onto clay tablets. But Napoleon would have been oblivious to these developments. He was a member of the Botai, a culture utterly and peculiarly obsessed with the horse. When Napoleon wasn't raising his penned and domesticated horses, he was hunting wild ones on foot. Horse meat was the bulk of his diet and horse milk made up the majority of his beverages. He would have drunk horse milk in the morning, then allowed it to ferment into the tangy alcoholic drink kumis and drunk it again in the evening. He tended to no crops or other animals with the exception of dogs. He used horse bones for his tools, horse hair for rope, and horse skin for leather; and when he died, his family would have buried him beside a horse.

If Napoleon ever rode horses before he invented his bridle, the experience was probably brief and done as a kind of rodeo stunt, according to Anthony. It may have been the kind of activity a risk-taking teenager without a fully developed prefrontal cortex might try. If he were born today, Napoleon would probably be familiar with the inside of a doctor's office. Psychologists would tell you he was overstimulated by novel situations. Everyone else might just call him an adrenaline junkie.

The horses Napoleon may have jumped on—though dangerous to ride—were already domesticated. They had, through thousands of years of artificial selection, become tamed and accustomed to humans. While scholars don't know precisely when, where, or how this process started, most paleozoologists now believe domestication was originally done accidentally, when cultures in the steppe penned wild horses near their camps in order to have a reliable supply of meat. Gradually, as the captors preferentially killed the wilder ones too difficult to manage and bred the tamer ones, the species became docile.

Domestication is extraordinarily rare in the animal kingdom. According to geographer Jared Diamond, it's only possible if an animal possesses six different behavioral and biological characteristics. First, he writes, the potential domesticate cannot compete with humans for food. It must eat scraps, like the pig, or, even better, eat something humans cannot, like grass. Second, a potential domesticate must also be able to breed in captivity. If, like the cheetah, the animal requires long mating runs or elaborate territorial displays, domestication is impractical. Third, the animal has to reach maturity quickly, so humans can efficiently raise and reap its benefits. Fourth, it has to be a pack animal with a social hierarchy, like dogs. Pack animals have a genetic predisposition for subservience, and humans can slide into the role of pack leader. Fifth, the animal cannot be skittish or possess a strong instinct to flee. This rules out animals like deer. Finally, unlike the zebra, the animal has to be either naturally docile or able to be bred into docility. The zebra's intractable aggression, regardless of selective breeding, has made it unviable as a potential domesticate despite what archaeologists can only assume have been repeated attempts.

The small number of successfully domesticated animals speaks to the extreme rarity of this combination. Humans haven't domesticated an animal of any significance in two thousand years (with all due respect to the ostrich), and the majority of the world's meat comes from only three domesticates. Even in the case of the horse, their domestication may have hinged on the discovery of a single unusually docile stallion. While mares naturally follow the leading stallion in a pack, wild stallions are hostile and fight with other stallions for females. They do not naturally follow, and are by default either searching for a pack of females or leading one. They would not have taken to the corral easily, and genetic tests on modern horses suggest that while many wild mares bred into the domesticated population, there may have been only a single "Adam" stallion. This stallion was presumably calmer, whose demeanor would likely have won him few successful couplings in the wild, but who apparently enjoyed wild success in the early horse pens.

By Napoleon's time, horses had been domesticated for untold generations. But with no means of controlling them, horses were simply a source of food and material before Napoleon invented the bridle.

The simplest form the first bridle would have taken is now called a "war bridle," which is nothing more than a leather rope looped around a horse's lower jaw and locked into place with a piece of wood. The ingenious invention can unobtrusively guide a horse thanks to a gap between their front incisors and rear premolars where a strap or, modernly, a metal bit can sit atop the gums. When tugged, the rope pinches the horse's gums, and the horse reflexively turns its head and body toward the tug to relieve the pain.

It's a simple tool that can be used to great effect. The Native American tribes of the Plains used war bridles without saddles and were considered some of the greatest riders in the world.

But the bridle's simplicity in design belies a complexity in concept. Napoleon may have been foolish with regard to his safety, but he was ingenious in regard to horses. There are no obvious parallels to the bridle found in the animals he may have been familiar with. The bridle couldn't have been ported over from cattle or sheep, as both have different mouth structures. How he came upon the idea is a matter of speculation. He likely tried placing a rope in all manners of locations to control his mount before he stumbled upon the unique gap between a horse's front incisors and premolars. Perhaps the most plausible explanation for Napoleon's discovery is simply a deep familiarity with the anatomy of horses. Maybe, even in a culture known for its horse obsession, Napoleon's stood out.

Napoleon's eureka moment is unsurpassed in the history of human transportation. By inventing brakes, Napoleon invented speed. With his well-placed rope, humans went from slow to fast, and horse riding remained the fastest way for humans to travel over land for more than five thousand years. A mounted rider wasn't beaten until August 28, 1830, when the steam-powered train Tom Thumb finally bested what Napoleon enabled.

Unsurprisingly, speed dramatically changed life on the steppes.

Anthony believes the Botai began to ride horses in order to hunt horses, vastly improving their productivity but also changing the structure of the culture. When the Spanish introduced domesticated horses to the Americas in the sixteenth century, they triggered a horseback riding arms race among the Native Americans of the Plains. As a trader in the American West wrote in 1851, "Men on foot cannot live, even in the best game countries, in the same camp with those who have horses. The latter reach the game, secure what they want, and drive it beyond the reach of the former."

The horse and its speed had another, more insidious effect for the villagers of the steppes. Before Napoleon's invention, the most dangerous part of a raid was the getaway. A raider could surprise his target and steal their goods, but the on-foot retreat from an armed and vengeful enemy posed a serious problem. With the bridle, horses became the world's first getaway vehicles, and thus tipped the balance of power in favor of the raider, says Anthony.

This had dangerous and damaging consequences. Without written records, it's difficult to confirm an increase in raiding in central Asia subsequent to Napoleon, but steppe villages added elaborate defensive walls in suspicious synchronization with the introduction of horseback riding. When the risk-taking Napoleon invented the bridle, he may also have invented a license to steal and life on the steppes may have taken an ominous turn.

At first, mounted warriors used their horses simply as transportation to and from battle and not in actual combat, largely because their weaponry didn't work on horseback. It took nearly a millennium for the Sintashta culture of the Russian steppes to invent cavalry by developing the chariot. Later innovations in the bow enabled Genghis Khan's riders to both ride and fire from horseback, which he used to rain terror over Asia. Horses continued to play a formidable role in battle for nearly six thousand years—from Napoleon's raids through Poland's cavalry charge at Krojanty in the beginning of World War II.

Yet perhaps the largest societal change the bridle wrought was the reallocation of resources and the rise of class, status, and stratified societies. Because a herder on horseback can control twice as many sheep or cattle as one on foot, in herding cultures, ownership, power, and wealth began to concentrate in a minority. After the invention of the bridle, archaeologists find a dramatic increase in elaborate grave goods—the leading archaeological indicator of income inequality.

Napoleon himself eventually found his way into one of these graves, perhaps by way of a tragically rapid deceleration. When Napoleon invented speed, he didn't just invent a new way to live—he invented a new way to die. And because evolution had not selected for hominins who feared speed as it had for those who feared snakes or heights, Napoleon was ill-equipped to under-stand the risks of his invention and dramatically unprepared for the new threat. It may be that Napoleon—the risk-taking, show-off genius—fell from his horse at full gallop. And the inventor of speed may have become one of the first to die from it.

11

Who Invented the Wheel?

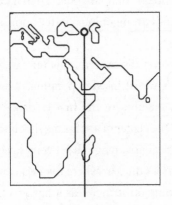

If the time of our species on earth
were a day, **this happened 25 minutes
before midnight** (5,400 years ago).

5,400 years ago

First wheel

Hundreds of thousands of years before the invention of the wheel,
some unlucky hominin stepped on a loose rock or unstable log

and—just before they cracked their skull—discovered that a round object reduces friction with the ground.

The inevitability of this moment of clarity explains the ancient ubiquity of rollers, which are simply logs put underneath heavy objects. The Egyptians and the Mesopotamians used them to build their pyramids and roll their heavy equipment, and the Polynesians to move the stone moai statues on Easter Island. But rollers aren't terribly efficient because they have to be replaced as they roll forward, and even if they're pinned underneath, friction makes them horribly difficult to move. The solution—and the stroke of brilliance—was the axle. Yet despite the roller's antiquity, it doesn't appear that anyone, anywhere, discovered the wheel and axle until an ingenious potter approximately six thousand years ago.

The oldest axle ever discovered is not on a wagon or cart, but instead on a potter's wheel in Mesopotamia. These may seem like simple machines, but they're the first evidence that anyone anywhere recognized the center of a spinning disk is stationary and used it to their mechanical advantage. It's a completely ingenious observation and so novel it's unclear where the idea came from—perhaps from a bead spinning on string?—as it has no obvious corollary in nature. The pole is called an axle, and many scholars consider it the greatest mechanical insight in the history of humankind.

Yet there exists another great intellectual leap between the potter's wheel and a set of wheels on a rolling object. The full wheel set appears to have first been invented by a mother or father potter, because the world's oldest axles are made of clay, are about two inches long, and sit beneath rolling animal figurines.

The first wheeled vehicle, in other words, was a toy.

As we've seen, there is a general reluctance in archaeology to describe any ancient artifact as a toy, but in this case the evidence feels overwhelming.

In July 1880, the archaeologist Désiré Charnay discovered the first pre-Columbian wheel set in the Americas. It was on a small coyote figure mounted on four wheels, and Charnay found it in the tomb of an Aztec child buried south of Mexico City.

As Charnay presumes in his book *The Ancient Cities of the New World*, the toy was a memento of "a fond mother . . . who, ages gone by, buried [it] with her beloved child."

The Aztec child lived thousands of years after the inventor in the high steppe but before the Europeans arrived with their wheels in the Americas, which suggests that in both the New and Old Worlds a mother or father potter independently invented the wheel and axle to make a toy.

The archaeologists I spoke with are hesitant to believe such a remarkable insight could have been made in the pursuit of something as frivolous as an object for play. The engineers, however, are not. Instead, they believe it would be far more remarkable if the first wheel and axles appeared on five-hundred-pound wagons. Small versions of inventions—modernly called models or prototypes—nearly always precede larger ones. They are far easier to build, take far less time, and allow an inventor to quickly discover potential problems and find solutions.

Yet as ingenious as this inventor was, their toy did not spark a societal revolution. The person who scaled it into a full-size set of wheels a few hundred years later did.

The full-size wagon first appeared approximately 5,400 years ago, and it may be the first invention in history to go viral. Archaeologists have discovered full-size carts from southern Iraq to Germany within a few hundred years of each other at a time when cultural barriers were particularly impermeable. The wagon, it seems, was irresistibly useful.

When I asked David Anthony, the anthropologist and author of *The Horse, the Wheel, and Language,* what explains this viral growth, he believes part of the reason may be the wagon's sheer size: "These were probably the biggest wooden machines anyone had ever seen," he says. They would have been loud; they would have been slow. And they were powered by teams of oxen, which were by themselves some of the largest animals in the steppe.

The invention of the wagon was the prehistoric equivalent of Sputnik; it did not go unnoticed. Because the two oldest wheels archaeologists have found vary significantly in design—one has an axle fixed to the wheel as it does on a modern train, the other spins freely on the axle like on a modern car—Anthony suggests that at least some wagon builders copied what they saw from afar without being able to inspect it closely.

The invention and widespread adoption of the wagon had an immediate and dramatic effect on societies throughout the Middle East and Europe. It drastically increased a farmer's productivity, and in doing so altered the landscape. Where once farms required teams of people to move the heavy loads of fertilizer, seeds, and crops, the wagon allowed for the possibility of a single-family venture. Populations that were previously clustered around rivers exploded onto the productive but unexploited steppe. The wagon changed entire economies, lifestyles, wars, and even languages. "It would be difficult to exaggerate the social and economic importance of the first wheeled transport," writes Anthony.

The wheel might have begun with a miniature, but the miniature didn't change the world. The full-size version did. And scaling a miniature wheel required its own genius. Anthony believes the full-size wheel and axle may have required craftsmanship so delicate, it was impossible to build with stone tools and could only be made with the gouges and awls metallurgists had only recently begun to cast. Finally, it couldn't have been made in stages, which means an individual built the first one.

The identity of the inventor of the first full-size wheel has become a cliché representing the unknowable, but recent reconstructions of long-dead languages have provided a powerful new piece of evidence that has brought scholars closer to the inventor than ever before.

Who made the first full-size wheeled vehicle?

I'll call him Kweklos—Kay for short—which paleolinguists believe is the word he may have used for "wheel." In his language, kweklos originates from the verb "to turn," meaning he referred to his invention very appropriately as "the turner." And I'll call Kay a him because the first known wagon driver is a man buried atop his wagon east of the Black Sea.

Kay was born roughly 5,400 years ago, a date that is well supported thanks to the popularity of Kay's invention. Wagons and references to them explode in the archaeological record from the

Middle East to Western Europe within a few generations of each other.

But if the when of Kay's birth has largely been established, the where is the subject of a lively academic debate. Anthony tells me the wagon "spread so fast that it's impossible to pinpoint a clear and obvious earliest date." As of now, two full-size wagon wheels tie for the oldest that archaeologists have found. One comes from a Slovenian bog in Ljubljana; the second comes from the remarkable Yamnayan culture grave just east of the Black Sea in the North Caucasus, Russia, where archaeologists found not only a wheel, but an entire wagon complete with the skeleton of a thirtysomething-year-old man sitting atop it.

Archaeology is not the proper science for pinpointing the location of viral inventions. There are, however, linguistic reasons to suspect the Yamnayan man buried with his wagon may have lived close to where the invention occurred. Many paleolinguists now believe the Yamnaya spoke a language called Proto-Indo-European (PIE), and reconstructions of that long-dead language suggest it's the native language of the wheel's inventor.

"The vocabulary for wheels shows that most of the words for wheels and axles were created by PIE speakers from their own verbs and nouns," Anthony tells me. For example, the PIE word for "axle" (*aks*) derived from the PIE word for "shoulder," which means PIE speakers used a word from their own language rather than a foreign one to refer to the wheel and wagon.

This fact is critical, because when cultures adopt foreign technology they typically adopt the originating culture's terminology as well. When the Spanish brought the tobacco plant back from the Caribbean, for example, they kept the local Taino word "*Tabako*." Reconstructions of wheel vocabulary suggest—though do not prove—that Kay was a PIE-speaking Yamnayan like the man buried with his wagon in the southwest corner of Russia.

Kay was a farmer and a herder. He had dogs, horses, and sheep, and perhaps wore some of the earliest wool clothing. He enjoyed mead, an alcoholic honey drink, and he raised cattle and drank their milk. He lived in a long house in a small farming community likely clustered near rivers.

Linguistic evidence suggests Kay worshipped a male sky god, sacrificed cows and horses in his honor, and lived in a village with respected chiefs and warriors. DNA of Yamnayans suggest Kay was likely to have had brown eyes, dark hair—though with an outside chance at red—and somewhat olive-toned skin. The average height for Yamnayan men was approximately five foot nine, and he likely had a heavily muscled frame from years spent toiling in his field.

Many of Kay's personal details are of course speculative, but one thing is certain: The builder of the first wagon was a genius at conception and craft. There is no other explanation. Scaling the toy into a full-size wagon meant solving a host of engineering, design, and woodworking problems. Some scholars, including Anthony, believe it isn't a coincidence that metallurgists first casted copper tools only a few generations before the first wagon. They believe the precise craftsmanship needed to construct a functional wheel and axle may have been impossible with stone tools.

The first and most critical component of the wheel, writes

Steven Vogel, author of *Why the Wheel Is Round*, is the fit with the axle. Too tight and the wagon is hopelessly inefficient, too loose and the wheel wobbles and breaks apart. The problem would not have been revealed by the matchbox-size wheel and axles, nor would models have required the proper ratio between the diameter and length of the axle. Too thick and the axle creates too much friction; too thin and it breaks under strain of the load.

Then there would have been the matter of the wheel itself, which is a deceivingly complex device. If Kay had cut a fallen tree salami-style for his wheel, it would have quickly failed. The problem, according to Vogel, is the direction of the grain, which in a salami-style slice of wood cannot support weight on its edge. Under strain, it would quickly deform. Kay's solution—as is evident from early wheel design—was to build a composite wheel out of multiple vertically cut planks. Kay would have had to carefully dowel these cuts together, and then shape them into a perfectly round wheel.

The size of Kay's wheels would also have been critical. Too small and the wagon cannot surmount any potholes, too large and the already heavy vehicle becomes immobile.

The genius of Kay's craftsmanship lies not in any single one of these realizations; it's in all of them. The wagon, as Anthony notes, could not have been put together in stages. It's all-or-nothing. Either it works, or it doesn't. But even with Kay's crafting genius, his wheel would have been uselessly massive were it not for oxen.

Cattle began as wild aurochs before the late Natufians of Turkey domesticated them some ten thousand years ago. At first the

Natufians used them exclusively for their meat and milk, but by the early fourth millennium B.C., the Maykop culture living in what is now Ukraine began to castrate the males and use them as work animals. The process of turning cattle into oxen and breaking them to the yoke would not have been pleasant. "It represented an entirely new level of domestication," writes the archaeologist Sabine Reinhold, "far beyond earlier intrusions into animal lifestyles." It involved castration, violence, and the infliction of pain. The animals "become lethargic," Reinhold writes. "Their spirits are broken."

It wasn't just the oxen who suffered. Tellingly, the early Yamnayan wagon driver who was uncovered by archaeologists suffered twenty-six separate bone fractures throughout his life, in addition to arthrosis in his spine, left ribs, and feet. His life was a brutal one, but perhaps not unusual. Many of the earliest Yamnayans buried with wagons show numerous broken bones, particularly on their hands and feet, probably because submitting oxen to the yoke was a violent struggle between man and animal. The Maykop buried their people with nose-ringed cattle, and some archaeologists speculate it was as a celebration of them having mastered the beast.

When Kay first yoked his oxen to his several-hundred-pound, three-by-six-foot wagon, complete with screeching wooden components and a team of oxen straining to haul it at a walking pace, he changed farming forever. Where it once took a village to haul a farm's heavy loads, with a wagon and a team of oxen it took only a family.

As a result, Yamnayan families, using their wagons as mobile homes, spread away from their clustered villages into the vast, unexploited Eurasian steppes—and then beyond.

Their cultural imprint is evidenced even today.

The Yamnayan rolled down from the high steppe into Europe and East Asia, bringing their wagons, culture, and language. Today, 45 percent of the world's population speaks a tongue descended from Kay's PIE. Languages as disparate as English, Greek, Latin, Sanskrit, Portuguese, Spanish, Swedish, Slovak, Pashto, Bulgarian, German, and Albanian, to name a few, can all trace their roots back to PIE.

Recent DNA evidence supports a similar conclusion: the Yamnaya culture moved from the steppes and swamped the cultures to its south and west. Their massive wagons played a large role in their cultural domination, but their smallest stowaway may have played an even larger one. Geneticists have located the bacterium *Yersinia pestis*, an ancient version of the germ responsible for the Black Death, in five-thousand-year-old teeth found in Central Russia and believe the Yamnaya may have unwittingly wielded a biological weapon as they rolled into Europe.

Perhaps the plague made Kay one of its early victims. Or, he may have died from an accident. The oldest known wagon driver has, after all, twenty-six separate bone fractures. Whichever it was, the early popularity of his invention is an indication Kay may have been one of the few ancient inventors to have been recognized within his lifetime. And because it became the funeral custom of the Yamnaya to bury drivers on their wagons, perhaps Kay was the first.

12

Who Was the Murderer in the First Murder Mystery?

If the time of our species on earth were a day, **this happened 25 minutes before midnight** (5,300 years ago).

5,300 years ago

First murder mystery

The predawn light had only just begun when the man watched his target set off from their Copper-Age village at the base of the Italian Alps. He watched him pick his way through the wooded foothills and waited for him to disappear up the mountain slope before he quietly gathered his equipment.

The man put on his sheep-hide undergarments and goatskin leggings and cinched his grass-and-sheepskin coat with a belt. He stuffed his leather and string boots with fresh hay, packed a meal of dried goat meat into his wood-framed backpack, tucked a flint knife into his belt, strapped a deerskin quiver of arrows to his back, slung his long bow over his shoulder, and set off in pursuit.

He was five and a half feet tall, thin and wiry, strong with legs and lungs well conditioned for the altitude and terrain. As he picked his way up the mountain, he did so with the speed and grace of someone who had spent a lifetime hiking the steep slopes. He kept pace with the man, but was in no rush to catch up. He knew the man would take the well-trod trading path at the top of the Ötztal Alps that the spring melt had only recently cleared.

He avoided the densely wooded Val Senales gorge, as his target had done, by climbing, descending, and climbing again, placing him to the south of what's now called the Similaun Pass. As he approached the ridge, he could see a wisp of smoke rising from a small gully up ahead. He unshouldered his bow, pulled an arrow from his quiver, and crept toward the top.

———

In September 1991, two trekkers in the Ötztal mountains hiking along the border between Austria and Italy stumbled upon a gruesome scene. The unusually warm summer had melted a patch of snow in a small trench along the Similaun Pass and exposed a dead man's torqued arm and grimacing face. The hikers assumed they had found the remains of someone caught in the last winter's storms and alerted the authorities. But as the medical examiners jackhammered him free, the man's freeze-dried skin led them to suspect he was far older—perhaps a lost soldier from World War I.

Then they spotted his equipment.

Lying beside the man, nearly perfectly preserved, was an ancient copper ax. It was a beautiful, prehistoric tool, rare even for its time, the likes of which no one had forged or wielded for nearly five thousand years. Archaeologists from around the world hurried to investigate, and local newspapers dubbed the man Ötzi, after the mountain range where he died. He had perished, they determined, some two thousand years before King Tut.

Ötzi's age by itself wasn't all that remarkable—archaeologists frequently discover far more ancient bones. The condition of his corpse, however, was unprecedented. A glacier had slid over his body immediately following his death like a freezer door sliding shut, sealing him into a small hollow for five thousand years. The result was near perfect preservation.

The chill and humidity preserved his DNA, skin, fingernails, organs, hair, eyeballs—even the contents of his stomach, all of which have provided scientists an unprecedented glimpse into the lives of Copper-Age Europeans. Today, after more than two decades of continuous examination, Ötzi's body has likely become the most studied corpse in human history.

Stanford archaeologist Patrick Hunt, a member of a National Geographic team who examined Ötzi, tells me scientists have determined his hair color, eye color, final meal, tattoos, arthritic joints, and numerous medical problems. By studying the enamel of his teeth, they know the precise valley where he was born. By reconstructing his vocal chords they know the sound of his voice, which was raspy and deep, a bit like a human bullfrog. And by the layers of pollen in his stomach scientists have re-created the exact route he took up the Ötztal mountains on his final trek.

But perhaps the most interesting revelation occurred nearly ten years after the hikers first discovered Ötzi, when a radiologist's scan spotted what others had missed: a flint arrowhead buried beneath his left collarbone that had left a severed subclavian artery in its wake. The injury would be a mortal wound on a modern operating table, let alone thousands of years ago on a mountain pass. The discovery settled ten years of debate about Ötzi's cause of death. He had not died of illness, exposure, or a fall. Ötzi's death was a homicide.

Violence has been a part of the human condition for far longer than we have been humans. Anthropologists once thought homicide and wars were a side effect of modern urbanization. Murder, went the theory, was the result of people living next to those they didn't know and wars were the result of the rise of state power. In recent years, however, this theory has been forcefully challenged, largely because it was based on two faulty observations: one, the idea that our ape ancestors were relatively peaceful and thus our evolution has been toward violence, and two, an underestimation of the violence in our ancient past. In recent decades, primatologists have discovered chimps kill each other with far more frequency than humans, and though hunter-gatherer battles result in fewer casualties than the organized slaughter of state warfare, their populations are far smaller and their violence can be chronic, resulting in equivalent or even higher rates of violence than anywhere but the most violent cities in the modern world. We have, anthropologists now believe, been killing each other with vigor since the very beginning.

What is new is the increasing taboo against violence. Most ancient killers received praise rather than punishment for their homicides. Yet as cooperation resulted in survival value and nicer people became selected for—a process some anthropologists call self-domestication—homicide became, at least situationally, taboo, and a killer could occasionally risk reprisal if their act was discovered.

Thus, the murder mystery was born.

Archaeologists don't know when this transition began, but Ötzi's murder is the first hard evidence archaeologists have uncovered of a killer attempting to get away with their act. After twenty years of investigating the most studied crime scene in human history, the evidence now suggests that Ötzi was assassinated by someone he knew who went to great lengths to cover up his crime.

Who killed Ötzi?

Let's call him Juvali, after the Juval district in northern Italy he likely called home, and a him because 96 percent of homicides around the world today are committed by men, according to a 2013 United Nations report. This is not a modern phenomenon.

According to the anthropologist Scott Brown, homicide as a predominantly male pursuit is a human universal that stretches deep into our past.

Juvali was born to farmers and herders rather than nomadic hunters and gatherers. His ancestors were pastoralists who originated in ancient Turkey and spread into Europe, slowly displacing the European hunters and gatherers through violence and disease, and bringing with them a settled lifestyle based upon tilled soil, domesticated animals, and permanent villages.

He may have had brown eyes and dark hair and perhaps sported a wispy beard, if he looked like a typical Neolithic European. Men averaged five foot five, which, when combined with his wiry strong frame, likely gave him the build of a modern-day horse jockey. The uneven diet that stunted his growth combined with his intense daily regimen would have kept his upper body lean but given him the powerful legs and degenerative joints of someone who had spent a lifetime hiking the steep Ötztal mountains.

Juvali likely suffered from poor health. The first farmers and herders in Europe experienced frequent bouts of malnutrition because—unlike hunters and gatherers, whose balanced and diversified diet protected them against the collapse of any one species or plant—Juvali's survival depended on a few crops and animals that could wither in a flood or drought. Even in good years, Juvali and his people ate a significantly poorer diet than hunters and gatherers, who were generally taller and healthier. Juvali also lived alongside domestic farm animals, which attracted pests, could foul his water supply, and provided an on-ramp to life-threatening viruses.

Serious disease was common for these early farmers. Many of their skeletons have deep grooves in their fingernails—called Beau's lines—that are indicative of illness. Juvali might well have suffered from tapeworms, arthritis, Lyme disease, or ulcers, all known to plague Neolithic Europeans. His starch-heavy diet likely meant he had a mouth full of cavities, and the steep terrain inflicted a heavy toll on his joints.

Juvali was not completely defenseless against the ravages of his unhealthy lifestyle. Recent discoveries have revealed these early farmers practiced far more sophisticated medicine than had been previously imagined. Juvali may have used antibacterial fungal remedies, Hunt tells me, and punched charcoal into his joints with a bone needle to ease his arthritis.

Researchers aren't sure of Juvali's profession, partly because they're unsure what kinds of professions existed in Copper-Age Europe. There were certainly shepherds, and farmers, and coppersmiths. But were there shopkeepers, barbers, or tailors? Twenty years ago, scholars believed the economies of Neolithic Europe were based largely on subsistence, but a close examination of Ötzi's shoes suggest a more sophisticated local economy than many scholars once presumed.

The shoes aren't much to look at. Through glass at the South Tyrol Museum of Archaeology in Bolzano, Italy, where I saw them, they appear like hollowed out loaves of bread held together with string. But soon after Ötzi's discovery they sparked the interest of Petr Hlaváček, a professor of shoe technology in Zlin, Czech Republic. He crafted exact replicas of Ötzi's string-and-leather, hay-lined size 7½ hiking boots and field-tested them. He slid on ice, dunked them in water, and wore them on his treks up the mountains of Europe. After his tests, he declared the shoes a "miracle." Their hay insulation kept his feet warm, even in snow, and their soft leather soles provided excellent traction with better pressure distribution than modern hiking boots. The Czech mountaineer Vaclav Patek wore a pair and proclaimed, "There is no mountain in Europe that couldn't be conquered in these shoes." Hlaváček believed no one but a professional cobbler could have crafted shoes of such quality, providing a hint, if not proof, of a potentially diverse array of professions in Juvali's village.

The sophistication of the shoes and potential existence of cobblers prove how little researchers still know about the kinds of trades that existed in Neolithic Europe, and make it difficult to guess Juvali's. Yet the most likely scenario is that Juvali was a herder.

Herding sheep and goats was (and still is) a primary means of making a living in the foothills of the Italian Alps. To this day, if you walk the Ötztal Alps, you will hear the gentle gonging of cow and sheep bells ringing along the hillsides. But it isn't merely playing the odds to suggest Juvali herded sheep or goats. His murderous disposition is also evidence of a herding lifestyle, because herders in ancient, lawless environments were a particularly violent lot.

Unlike farmers, whose wealth is embedded within the ground, a herder's life savings sits atop four legs, which leaves it exposed to thieves. Before modern law enforcement, a herder's only defense against robbery was the promise of violent retaliation. Without fair courts to settle disputes, the incentive for a herder was to overrespond to challenges, threats, and thievery.

The results of these perverse incentives can be seen even in relatively recent history. In medieval England, when the records office was far better than the court system, the average herding hamlet suffered homicide rates that rivaled a modern-day war zone. In 1340, the herders and farmers of Oxford, England, slaughtered each other to the tune of 110 killings per 100,000 citizens. Today, the rate is less than 1 percent of that.

Archaeological anecdotes hint of even more extreme violence in Juvali's era. Many of the world's most famous archaeological discoveries contain grisly footnotes—smashed skulls, embedded arrowheads, cut throats, and signs of strangulation. One of the oldest skeletons in North America is the 9,400-year-old Kennewick Man found along the banks of the Columbia River. He also happens to have a stone spear point buried in his pelvis. The Lindow Man is a 2,000-year-old, well-preserved body found in an English peat bog. He died from a smashed skull, a cut throat, and a broken neck. The Tollund Man is a 2,500-year-old body found in Denmark so perfectly preserved in a bog that you can read his final expression and see the noose that hanged him tied around his neck. A Swedish cemetery dating from this era was excavated and the lead archaeologist, T. Douglas Price, determined that 10 percent of the deaths were violent. The modern rate is 1.4 out of 100,000. In Juvali's herding village, the evidence suggests his homicide would not have been a particularly unusual occurrence.

Recently, the South Tyrol Museum of Archaeology hired the

Munich homicide detective Alexander Horn to investigate the Ötzi case. As in any homicide, he first tried to establish Juvali's motive. Almost immediately, he ruled out robbery. The most powerful piece of evidence at the scene is the presence of Ötzi's rare copper ax, which is like finding a Rolex on the wrist of a modern homicide victim. Not only can robbery be ruled out, but if a member of a rival clan had shot Ötzi they would have taken the ax as a trophy. The best explanation, according to Horn, is that the murderer left the ax because they had common acquaintances who would have recognized it. In other words, Horn believes Juvali knew his victim, which is—even if we were simply guessing—by far the most likely scenario.

According to the University of Virginia sociologist Donald Black, fewer than 10 percent of modern homicides are committed for practical gain such as robbery. Most homicides, he writes, are actually cases of capital punishment in which the victim is tried, convicted, sentenced, and executed by a killer who feels justified. Extending this modern analysis of homicide to one committed so long ago may feel like a leap, but a bone-deep gash on Ötzi's hand is further evidence Juvali might well have been set on revenge.

The cut is a classic defensive wound, according to Horn, and the degree of healing on the wound suggests the injury occurred some forty-eight hours prior to Ötzi's death. Hunt also tells me Ötzi's own flint knife has a broken tip that would have rendered it useless. The fact that he hadn't repaired it yet suggests he snapped it shortly before his death. Finally, investigators have identified someone else's blood on Ötzi's cloak. Taken together the blood, the knife, and the wound all suggest Ötzi wounded or perhaps killed someone twenty-four to forty-eight hours before his own death. Perhaps, Horn suggests, Juvali knew the victim, providing him with a powerful and timeless motive as he tracked his target up the mountain.

The weather must have been pleasant the day Juvali shot Ötzi. Carbon embers on the bacon and bread found in Ötzi's stomach suggest he built a fire in a small gully along the mountain pass and casually ate his lunch. By the food's position in Ötzi's digestive system, researchers estimate this final meal occurred approximately a half hour before his death.

Ballistic analysis indicates Juvali took a position approximately one hundred feet from Ötzi and fired his arrow from just below him and to his left. One hundred feet is roughly the length of a basketball court, and considering his prehistoric equipment, it was a phenomenal distance. Even modern bowhunters shooting carbon fiber arrows at deer prefer to be closer. But Juvali's aim was true. His arrow pierced Ötzi's left shoulder blade, severed the main artery that provided blood to his arm, and stopped just below his collar bone. He died within minutes from arterial blood loss.

Juvali then approached, surely knowing his arrow had found its mark. He flipped Ötzi over as he lay dying—explaining the corpse's oddly torqued shoulder—and wrenched his arrow from his back, separating arrowhead from shaft. Archaeologists excavating the scene found fourteen arrows, but they all belonged to Ötzi. Juvali's arrow is gone. Juvali likely removed the identifying shaft, Horn believes, for the same reason a modern assassin picks up bullet casings. He may even have buried his victim in snow—which would help explain Ötzi's remarkable condition.

The lengths Juvali went to in order to cover up his crime are in the end the most powerful pieces of evidence against him. If Juvali had wanted to simply kill Ötzi without witnesses, he didn't need to follow him to the top of a mountain. Doing so suggests not only that he didn't want witnesses, but that he wanted to obscure Ötzi's cause of death. By following him into the Similaun Pass, his actions suggest a motive so obvious that if the townsfolk knew Ötzi's true cause of death, Juvali would have been the prime suspect. To get away with murder, he followed Ötzi up the Alps, believing the remoteness of the location would prevent the body's discovery.

In this, he turned out to be spectacularly correct.

13

Who Was the First Person
Whose Name We Know?

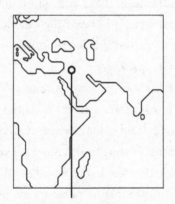

If the time of our species on earth
were a day, **this happened 24 minutes
before midnight** (5,000 years ago).

5,000 years ago

First name we know

On November 24, 1974, the archaeologist Donald Johanson
spotted a tiny bone in a small gully near Hadar, Ethiopia, that
added a critical branch to the ancestral tree of our species.

The bone belonged to a 3.2-million-year-old female of a new species, neither human nor ape, later dubbed *Australopithecus afarensis*. Over the next several months, Johanson and his team uncovered nearly half of her skeleton. This at a time when, Johanson later remarked, "most of the evidence for human evolution older than 3 million years could fit in the palm of your hand." The find stunned the archaeological world, but it also captured the public's imagination. The ancient hominin soon became prehistory's greatest celebrity. Perhaps its *only* celebrity. And that's largely, according to Johanson, because the archaeologists gave the bones a name: Lucy.

The name choice was random, taken from a Beatles song playing at the team's camp that night. But it imbued the ancient skeleton with humanity and closed the emotional distance between modern readers and this long-lost relative. Besides, to be famous, you must have a name.

The naming of Lucy is the most famous example of what archaeologists are always forced to do with the many people or places who lived or existed before writing. Of course, we have no idea what Lucy's friends and family called her when they wanted her attention or whether she had a name at all. Nor do scholars know the name of virtually any person, city, or culture beyond the past six thousand years. The names scholars use today for these people and places are a completely modern fiction. Sometimes they're based upon the local geography; sometimes on the Beatles song playing on their radio.

It's estimated that roughly 1.5 billion *Homo sapiens* lived and died before the invention of writing. And not a single one of their names is known or will ever be known to anyone. Nor are the names known of many who lived after the invention, when writing was used by only a few people living in a few places. We know what five-thousand-year-old Ötzi the Iceman ate for his final meal (see chapter 12), but we don't know his real name. Nor do we know the name of the first person to set foot in America, or the names of the great painters of the Chauvet or Lascaux caves (nor even what the painters called those caves). Not until the rise of the Mesopotamian civilizations and the invention of writing do we finally learn the given name of an individual. And he wasn't a conqueror, a prince, a queen, or a king. He was an accountant, which is entirely appropriate. Unlike kings or conquerors, accountants are the reason any name is remembered, because while they're not often thought of as pursuing a creative vocation, they're responsible for what most scholars believe is humanity's most creative invention.

Writing did not emerge in a single moment of eureka! Instead, it first evolved in Mesopotamia over nearly five millennia as a solution to an increasingly complex concern: separating what was mine, what was yours, and—most important—what you owed me. A few thousand years later, the Incans independently developed an entirely different system of writing called quipu, based upon strings and knots, but appear to have done so for exactly the same reason: to record debts and ownership.

In Mesopotamia, writing began with tokens.

In the 1960s, as excavations of ancient Mesopotamian cities began in earnest, archaeologists uncovered thousands of small pieces of shaped clay. These tokens came in a variety of designs.

Some looked like animal heads while others were round with marks scratched into them. Some looked like charms, others like a potter's absentminded creations. Archaeologists were puzzled. No one knew what they were or their importance, and archaeological reports from the era simply note their existence. In some cases, they were even thrown away. That is until the 1980s, when the historians Denise Schmandt-Besserat and Pierre Amiet came to a startling conclusion: the tokens weren't art, or charms, or absentminded clay doodles. They represented the first step toward the invention of writing.

The tokens, according to Schmandt-Besserat, worked something like this: Imagine you lived in ancient Mesopotamia nine thousand years ago and you herded cattle. Trade existed, but it was relatively straightforward. If a neighbor wanted one of your goats but didn't have something in return, you remembered their debt as you would remember a friend's debt today. Simple enough.

But as the population of Mesopotamia grew and coalesced, transactions increased, and trade grew too complex to remember who owed you what. For the first time in history, business outgrew the brain's capacity. As a result, Mesopotamians began using a simple yet ingenious system of tokens. If someone took one of your cattle, they would give you a small clay token shaped into cattle's horns in exchange. That way, when you wanted, say, one of their sheep, you could cash in their token. These tokens sound a bit like money—but they aren't. At least not yet. They have yet to reach that level of abstraction, though one can see how a system of exchanging debts would eventually spawn money. After all, if you remove the abstraction, money today is a measure of a government's debt to you. And when you want a Coke, you can give the store clerk the physical representation of that debt in exchange.

As trade grew in its complexity, a system of scattered tokens grew untenable, and beginning approximately six thousand years ago Mesopotamian traders began sealing the tokens into hollow clay balls. These clay balls seem to have functioned as the world's first filing cabinets, allowing different transactions to be compartmentalized from one another. But filing cabinets require labels, and so herders impressed the images of the tokens onto the outside of the clay balls. This seemingly simple and obvious step actually represents an important middle ground between physical representations and abstract symbols.

Of course, a drawer of jumbled balls is no way to run a business, and Mesopotamians again transitioned to a higher level of symbolism. They began impressing the balls and their symbols onto clay tablets, separating each transaction into their own boxes on the tablet, which in effect created the first rules of grammar. These tablets could pack far more information into a far easier to read format. Best of all, they didn't roll away. From there it didn't take long before Mesopotamian accountants abandoned the balls and tokens entirely and simply inscribed their symbols onto the clay tablets.

Soon after this development, the accountants added their identifying symbol onto the spreadsheets. When archaeologists found one of these spreadsheets, translated the accountant's marking at the bottom, and dated it as the oldest ever found, this accountant became the first person whose name archaeologists don't have to make up based on a Beatles song or a mountain or a local village. Instead, a person from the past speaks with us directly.

His name was Kushim.

Kushim was born five thousand years ago in ancient Mesopotamia, in what is today southern Iraq, a few hundred years before the first Egyptian hieroglyphs. Kushim doesn't tell us his sex, but because the earliest Mesopotamian proverbs are largely written from the male perspective, according to the Assyriologist Bendt Alster, the consensus is that most of their scribes were men.

Kushim lived along the banks of the Euphrates River just north of the Persian Gulf, where the Euphrates and Tigris nearly converge before they pour into the Red Sea. The two rivers brought with them fertile soil into the valleys, and their ancient paths and periodic floods formed some of the most productive agricultural plains in the world. These lands and the crops they hosted fed some of the first great cities, including the largest in the world at the time: Uruk, where Kushim lived.

Uruk packed an estimated fifty thousand people into dense mud brick housing and served as a trade hub throughout the Mesopotamian region. The city had an aristocracy, large temples that distributed grain and beer, and schools that taught a remarkably homogeneous curriculum. Archaeologists have found 165 separate copies of the same writing assignment spread out over a thousand years. It's called the list of professions, and it was a sort of standardized test in Mesopotamian schools, passed down among scribal schoolteachers for a period of time nearly three times as long as the United States has existed.

Kushim's literacy suggests he attended one of these schools, which in turn suggests he wasn't born enslaved. This already puts him in an exclusive group in Uruk. Enslaved people formed a significant fraction of the population and a bulk of the workforce. A single textile factory in Uruk enslaved thousands, and enslaved people almost certainly dug Uruk's extensive irrigation channels. Much of the early writing from Mesopotamia deals with slave ownership. After Kushim, the next names history remembers are the slaver Gal-Sal and the two people he enslaved, named Enpap-X and Sukkalgir. In Sumerian, the word for "slave" is somewhat ominously a synonym for "foreigner."

As a child, Kushim shared experiences familiar to modern schoolchildren. Around the age of five or six he would have begun attending scribal school, which was described in later writings by other students as both "long and tedious." A composition called "Son of the Tablet House," written by a student nearly one thousand years after Kushim, described a typical day as: "I read my tablet, ate my breakfast, prepared my new tablet, wrote and finished the text. Then in the afternoon I was given a recitation."

If Kushim was late to class his teachers expected him to bow in apology. If he was lucky, they wouldn't beat him. Usually he wasn't lucky. Most transgressions, however slight, resulted in beatings. One ancient Sumerian proverb goes on at length describing the various reasons a student like Kushim might find himself on the wrong end of a cane, including not standing when the teacher arrived, leaving without permission, and poor handwriting.

To make amends, Kushim might invite his teacher (scribal teachers were more often male, but not always) to his home, ask him to sit in the seat of honor, and have his parents ply the teacher with a new robe or new ring and a meal. If the teacher deemed the apology and/or bribes sufficient, he might, according to "Son of the Tablet House," say something like: "Since you did not show disrespect for my words, did not ignore them, you will climb to the top of the scribal arts . . . may you be the leader among your brothers, and among your friends . . . you have accomplished your duties as a pupil, you are now a cultured man."

In school, Kushim learned to make his own clay tablets and bake them in the sun. Archaeologists occasionally find ancient stacks of blank tablets, which they assume represent early class assignments for students like Kushim. Clay tablets, once dried in the sun, are virtually indestructible, which is how archaeologists today have been able to draw such a clear picture of daily life and particularly of the classwork in Uruk.

To learn to write, Kushim learned to copy sign elements like a modern student practicing the letter "E" by tracing the various pencil strokes that comprise it. Archaeologists have found tablets with partial signs copied over and over again that look like tedious ancient worksheets, which is emblematic of an education that was both long and difficult. Cuneiform script had nearly one thousand different signs, all of which he had to memorize. And then there was the problem of numbers: No one had invented them yet. For Kushim, sheep and goats and years all had their own counting systems, because no one had observed five people, five sheep, and five fingers and realized they shared this numerical abstraction we now call "five" in common. Numbers are so ingrained it's difficult to realize what an ingenious and nonobvious invention this was (you can begin by trying to define the number "five"), but numbers allow us to compare, contrast, and measure radically different concepts. Unfortunately for Kushim, that genius lived after he did, which drastically increased the amount of signs he had to learn. Scholars believe his education continued into adulthood.

Once he graduated, Kushim occupied a relatively lofty position as a temple administrator responsible for brewing large quantities of beer. To keep track of his duties, Kushim practiced a form of protowriting based upon pictographs. His sign for barley looks like a stalk of barley, and you can still see the remnants of the old cattle's horns token in his representation for cow. Kushim couldn't write a love poem or story, because his pictograph system didn't have a complete grammar, nor could it represent many sounds. When Kushim wanted to identify himself, he didn't have a suite of supple letters to do so. His solution was to put the two pictograph signs "Ku" and "Sim" together in the same way that if your name were "Carpet," and you didn't have letters, you might sign:

The word "Kushim" itself is merely a bit of poetic license by the Assyriologist Hans Nissen, who put a sound to the "Ku" and "Sim" signs, because for Kushim, writing may have had no phonetic value. Whoever read the debt may have simply treated his name as we would a family crest.

As an individual, Kushim might have been, somewhat ironically, quite forgettable. He appears to have been so extraordinarily normal he may even have owned the world's first mass-produced item: a beveled-rim bowl, an item that archaeologists have found everywhere in Uruk and which, when filled with grain, seems to have been the ancient version of a paycheck. At the end of his workweek, Kushim's employer would fill his bowl with an allotment of grain to compensate him for his service.

Perhaps as the administrator of a large temple he was well "paid." On the other hand, he wasn't a very good accountant. On one of his tablets he erroneously writes "10" instead of "1"—mistakenly ordering ten times the necessary amount of barley. On another, his recipe for beer doesn't call for enough malt, which is a particularly egregious error since the ratio of barley to malt should have been the very simple 1:1.

But what he lacked in attention to detail, he made up for with what Nissen writes is "a bureaucrat's zeal for exaggerated accuracy." In one instance, Kushim calculated 135,000 barley liters (81 tons) to the nearest five. A "painstaking accuracy that stands in complete contradiction with the numerous mathematical errors," according to Nissen.

Kushim—the first person whose name we know—was a normal, bureaucratic, error-prone administrator charged with tallying debts. And that's entirely appropriate.

For a thousand years, as scribes and accountants developed writing in Mesopotamia, it was the exclusive domain of the aristocrats, bureaucrats, and administrators like Kushim. The general populace either ignored it or, because tax collectors wielded it like a weapon against them, despised it. Tellingly, when Mesopotamian cities fell to disease or invaders as they often did, the citizens nearly always took their torches to the library first. Common citizens felt the same way about writing as a modern reader probably feels about the tax code, because that's all the entire medium represented. It was the exclusive domain of the Kushims and the Arthur Andersens, not the William Shakespeares.

When Kushim died, it may not have been to much acknowledgment, given his station in society and the disdain for his vocation among the general populace. But if he had a headstone, it could at least have been among the first with a name written upon it.

14

Who Discovered Soap?

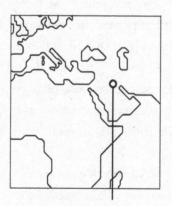

If the time of our species on earth
were a day, **this happened 22 minutes
before midnight** (4,500 years ago).

4,500 years ago

First soap

On the morning of March 14, 1942, Anne Sheafe Miller lay
dying on a hospital bed in New Haven, Connecticut.

A month earlier the thirty-three-year-old nurse had suffered a miscarriage and contracted a common infection. Despite state-of-the-art medical care, including blood transfusions and sulfa drugs, her temperature gradually climbed to 107 degrees and she began slipping out of consciousness.

"She had, on the basis of all previous experience, fatal hemolytic streptococcus septicemia," her doctor John Bumstead wrote in his notes. Desperate, Dr. Bumstead procured an experimental drug from a lab in New Jersey. The drug was so rare that technicians would later recycle it from Miller's urine.

A state trooper escorted the coarse, dark brown powder to Miller's bedside in New Haven, where the attending bacteriologist, Dr. Morris Tager, examined the foul-smelling drug with "some concern and suspicion," he later wrote. Nevertheless, Dr. Tager injected it into Miller that afternoon. By the next morning her temperature had returned to normal, and Miller went on to outlive her initial prognosis by some fifty-seven years.

The experimental drug is now known as penicillin, and Anne Sheafe Miller's life is the first of more than 200 million it has since saved.

Penicillin was the world's first antibiotic—a class of medications that has revolutionized humanity's fight against infections—but antibiotics are only humankind's second most effective weapon in our eternal struggle against malignant bacteria. No medical product, and perhaps no medical discovery, has saved more human life than soap.

Ever since the Dutch glassmaker Antoine Philips van Leeuwenhoek peered through his microscope in 1668 and saw a living organism in his lens, we have been slowly coming to terms with just how many bacteria inhabit our world. Earth is so infested that if aliens ever visit they'll likely declare our planet a bedbug infested motel and promptly leave. Bacteria are on the pillow we sleep on at night and on the cereal spoon we use in the morning. Biologists estimate a single human hand plays host to 150 different species at any given moment. Most are harmless, some are beneficial, and some are murderous if they could only find a way past the skin. In densely populated cities, where these dangerous bacteria and diseases can be passed from one to hundreds with a twist of a doorknob, it's no exaggeration to say soap doesn't just make cities healthier—it makes them possible.

Thanks to its antiquity we will likely never know exactly how many lives soap has saved—the researchers at the Science Heroes website that tracks these things threw up their hands when I posed the question to them—but even the conservative estimates run into the hundreds of millions. And yet it could save far more. According to a report by UNICEF, if every cook used soap it would cut the world's rate of respiratory infections by 25 percent and cut diarrheal diseases in half. That alone would save more than a million lives every year.

Soap is serially underestimated and underutilized because it suffers from a fundamental PR problem: it cleans away something you cannot see. That's a difficult conceptual leap for even the most educated among us. According to the US Centers for Disease Control and Protection, doctors wash their hands half as frequently as they should. Soap also saves the life of a healthy person oblivious to the bullet they dodged—while an antibiotic can take someone like Miller from death's door to the dance floor overnight. So despite not only saving more lives than penicillin but enabling our modern urban existence, we still underrate what is perhaps the greatest medical discovery in human history.

Who discovered soap's recipe?

I'll call her Nini, after the Sumerian goddess of medicine Ninisina. And I'll call her a "she" because soap's discoverer likely worked in Sumeria's burgeoning textile industry, which the anthropologist Joy McCorriston tells me was an industry dominated by women.

Nini was born 4,500 years ago in what is now southern Iraq, perhaps in the ancient Sumerian city of Girsu, where the oldest written tablet detailing the manufacture of soap originates. Girsu was one of the world's first cities with a population in the tens of thousands and was at one point the capital of the ancient Lagash kingdom. It is home, among many other of humankind's firsts, to the world's oldest known bridge.

Nini was born around the time of the construction of the Great Pyramid of Giza, and other than being slightly shorter than the average person today was completely modern in appearance. In their cuneiform tablets, Sumerians often referred to themselves as the "dark-haired people." Though many Sumerian statues feature big blue eyes, most historians believe this was a feature of their gods and that blue-eyed Sumerians were actually quite rare. Nini's eyes, like her hair, were probably dark. Pictographs suggest Nini likely wore garments of fringed fleece or sheepskin that fell to her ankles, which she embellished and decorated with patterns, while men wore kilts and sashes.

Nini grew up in a bleakly patriarchal society according to the Mesopotamian scholar and author of *Women in Ancient Mesopotamia*, Karen Nemet-Nejat. Her father was the head of her household and exercised authority over her until he died or she was married, which may have happened when she was a teenager, or sometimes even younger. One Akkadian text describes a wife as three feet tall at the time of her wedding.

Marriage for Nini was a contract between her father and father-in-law. Sumerians viewed a woman's marriage as less to her groom than to his family, so if Nini's husband should die she would likely be remarried to one of his brothers. Women in Sumeria had virtually no prospects for independence, according to Nemet-Nejat. And at no time in Nini's life would she have been independent of either her father, her father-in-law, or her husband.

Nini's primary role as a wife was to bear children, especially sons, who were viewed as heirs. Marriage was traditionally monogamous, but if she did not mother children a husband might take a second wife or a concubine, or adopt. In a hymn, the goddess Eula describes a Sumerian woman's life in simple terms:

I am a daughter,
I am a bride,
I am a spouse,
I am a housekeeper.

Nini likely grew up lower class because in addition to being a housekeeper she had a far more modern role. The Mesopotamians, among their many gifts to humanity, invented the finger-numbing drudgery of factory labor in textile mills. These large state-run industries depended upon enslaved labor, debtors, and quasi-employees to shear, sew, dye, and produce woolen textiles, a critical export for many Mesopotamian cities.

By the time of Nini's employ, Girsu's textile factories were large-scale production centers impressive even by modern standards. The archaeologist Daniel Potts calculates that in a three-month period, 203,310 sheep were shorn in Girsu alone, a number made even more impressive since it was achieved before the invention of sheep shears. Instead of efficiently cutting their wool, sheep-shearers plucked each individual strand, one-by-one, from all 200,000 sheep. Because of this painstaking pace, Potts estimates that a single weaver only produced an average of ten inches of wool per day. Yet according to the Assyriologist Benjamin Studevent-Hickman, thanks to as many as ten thousand laborers a single textile factory in Ur produced more than four hundred tons of wool in a single year. Nini, it seems, was one of these workers.

She may even have occupied a position of some authority. Female supervisors typically managed thirty-person teams in these factories, according to archaeologist Rita Wright, and received higher wages in the form of barley or cloth. Given Nini's influence in introducing soap to the manufacturing process, perhaps she was one of these supervisors.

The first documented use of soap is described on a cuneiform tablet found in Girsu. According to chemical archaeologist Martin Levy, the tablet was written 4,500 years ago and concerns the washing and dyeing of wool. To properly dye wool, a weaver must remove the lanolin fats from the textiles, which is accomplished far more easily with soap. Even today, weavers wash freshly sheared wool in soapy water to remove the lanolin.

Nini was far from the first person to take advantage of the chemical reaction between alkalis and fats—what chemists call saponification. The ingredients are common enough that most scholars suspect someone long before Nini first created the reaction accidentally, according to Seth Rasmussen, a professor of chemistry at North Dakota State University with whom I spoke. Alkalis are found in the ashes of burned wood and many scholars believe early humans used wet ash to clean greasy butchering tools. Unbeknownst to the cleaner, ash combined with the animal grease to create a simple, impure soap.

The fact that wet ash removed grease was probably understood by the first weavers as well, who likely used it to clean their textiles, according to Hugh Salzberg, author of *From Caveman to Chemist*. The ash would have combined with wool's lanolin to create saponification.

Yet there's reason to believe no one discovered one could make soap itself—which one could then wash one's hands with—before 5,000 years ago, according to Rasmussen. Because there's no mention of soap for the first millennia of Mesopotamian writing, most scholars believe soap was discovered in some proximity to its first mention in the tablets 4,500 years ago. "If it was known long before Sumerian times we would expect to see references to it in earlier Sumerian records, which we don't," Rasmussen tells me.

Nini's stroke of genius was likely the moment she discovered the fatty lanolin or the animal grease was the reason ash worked so well as a cleaning agent, and could be added to ashy water to create a bucket of liquid soap. It may seem a small step, but it meant Nini no longer relied on the grease of whatever she was washing to aid in the reaction. Instead, she could create the ideal mix of fats and alkalis and wash anything—especially, and most critically, human hands.

Nini's first soap, Salzberg theorizes, may simply have been a bucket of ashy, greasy water. Later, either Nini or someone else realized they could strain the ash and globules of fat away, as water absorbs the alkali from the ash in a process called "leaching." Because few people would have been interested in washing themselves with ashy water, leaching was another important step in encouraging people to use soap for its most useful purpose. Eventually, and up through the Middle Ages, soap makers skipped the strainer and dipped sacks of ash into water like tea bags.

The first known recipe for soap calls for approximately one quart of oil and six quarts of potash (potassium leeched from wood ash). According to Rasmussen this would have combined to create an impure but useful liquid soap. Using this crude formula to produce her ashy, greasy water, Nini would have made the most lifesaving medical product ever developed by humankind.

Except she wouldn't have known it.

Soap is not a bacteria destroyer, like penicillin. Instead, it is a bacteria remover. By creating molecules that attach to both oil and water, soap creates what chemists describe as an "oil suitcase," allowing water to briefly coexist with oil. Any bacteria hiding within or beneath it can then be washed away. Because the lifesaving effect of soap is so difficult to observe, it's unlikely Nini received much recognition or acknowledgment in her lifetime nor would she have had any idea what she had done. Perhaps she received some recognition at the textile plant for her slightly more efficient method for cleaning off lanolin. An increase in her allotment of barley, perhaps.

Sumerians almost certainly didn't wash their hands with soap for the same reason modern doctors often fail to—because their hands already look clean. For hundreds of years after its invention there's no evidence anyone used soap to clean their bodies. Instead it was used on items like dishes or clothing that had obvious grease stains.

The first proof that anyone used soap to clean their skin comes from a cuneiform tablet found in the Hittite capital of Boghazkoi and written nearly a thousand years after Nini. The tablet reads:

With water I bathed myself.
With soda I cleansed myself.
With soda from a shiny basin I purified myself.
With pure oil from the basin I beautified myself.
With the dress of heavenly kingship I clothed myself.

All technologies, no matter how significant a breakthrough they represent, take time to spread throughout a population. Economists call this lag "technological diffusion," and Nini's discovery suffered a particularly elongated one. Global adoption of soap is, as we've seen, still under way and its five thousand year progression has hardly been linear.

Beyond Mesopotamia, soap's adoption was even more haphazard and often neglected for objectively worse alternatives. Instead of soap, Roman launderers used putrefied urine collected from passersby. Pliny the Elder credited soap's discovery to the Gauls (the "barbarians"), who used soap instead of urine. The Greeks primarily washed themselves with soapless water even though Galen, the Greek physician, presciently recommended its use as a preventive medical treatment.

The failure to understand soap's power to save lives persisted not only because bacteria were invisible, but because they were drastically misunderstood. Even as late as the 1800s, the mere suggestion that microorganisms might be lethal could get one fired from one of the most advanced hospitals in the world.

In 1847 Dr. Ignaz Semmelweis, a Hungarian obstetrician working at the Vienna General Hospital, began investigating a mystery. Why did birthing mothers under the care of his doctors die at a rate five times greater than by those cared for by only his midwives? Semmelweis insisted his doctors follow every one of his midwife's birthing techniques. He even went so far as to ban priests from ringing bells in the doctor's wards because they didn't in the midwives'. Nothing worked. But when a colleague of Semmelweis's died after performing an autopsy of the same maternal fever that plagued mothers, he tried a new theory. He knew his doctors often performed autopsies before they assisted mothers, while his midwives did not. Perhaps, he reasoned, they carried some deadly particle or smell from cadavers to mothers, and thus he asked them to wash their hands. The death rate immediately plunged, but Semmelweis's doctors quickly mutinied at the suggestion that they were responsible for killing their patients. He was fired, the doctors stopped washing their hands, and Semmelweis eventually perished in an insane asylum.

Nini, the Assyrian textile worker born thousands of years before him, probably suffered a similarly anonymous passing. Average life expectancy in ancient Sumeria is estimated at only forty years, in part because early cities were cesspools of disease and infection. As Nini lived amid a dense population with not a soul who washed their hands with soap, she may have died from the very kind of bacterial infection her discovery saves millions from today.

15

Who Caught the First Case of Smallpox?

If the time of our species on earth were a day, **this happened 19 minutes before midnight** (4,000 years ago).

4,000 years ago

First case of smallpox

The greatest killer in history is not Genghis Khan. Or Alexander the Great. Nor is it any one battle, war, or even all wars *combined*.

Rather, it's a single, hapless, blameless individual, born four thousand years ago in the Horn of Africa, who just so happened to breathe in a novel virus hitching a ride on a speck of dust.

Variola—the virus responsible for smallpox—nearly wiped out the New World and took a solid stab at the old one. In the eighteenth century, it killed four hundred thousand Europeans every year. In the twentieth century, it killed three times as many people as both world wars combined, and it played a significant role in the fall of the Aztec and Inca empires after European sailors introduced it.

Variola had no particular predilections. It killed young and old, kings and peasants. The Roman emperor Marcus Aurelius died of smallpox, as did Queen Mary II of England, Tsar Peter II of Russia, and King Louis XV of France. Abraham Lincoln came down with smallpox a week after delivering the Gettysburg Address.

Smallpox decided wars. In the Franco-Russian conflict, the Germans were vaccinated, the French were not, and nearly half of the French casualties were due to smallpox. Some scholars have called George Washington's order to inoculate all his troops his most important tactical decision of the Revolutionary War.

Unlike the Black Plague, smallpox isn't a disease modern medicine has long conquered. If you were to come down with a case tomorrow, a hospital wouldn't improve your chances of survival any more than a witch doctor. And if someone distributed *variola* over a modern city, thousands, perhaps millions, would die. In 1972 a single clergyman returning from Mecca sparked an outbreak in Yugoslavia that killed 35 people and sickened 175 before roadblocks, house-to-house searches, curfews, and a declaration of martial law halted the spread. Smallpox is the greatest viral scourge in the history of humankind, and it has left a heavy footprint on our species.

In 1959 the World Health Organization, armed with a vaccine, launched a campaign under the direction of the American epidemiologist D. A. Henderson to eliminate smallpox. Because the virus can only infect humans and must infect a new person every fourteen days to survive, its eradication was at least theoretically possible—if absurdly ambitious. Wiping a human virus off the planet had never been done before, and for good reason. Three million viral particles could fit comfortably within the period that concludes this sentence, and in the year the program began, smallpox had infected at least 15 million people. And yet on October 26, 1977, Henderson and his WHO team had lowered that number to one. The global hunt had isolated the last of *variola* within the body of a single person—a Somalian hospital cook named Ali Maow Maalin.

Maalin was a twenty-three-year-old man who lived in the town of Merca, Somalia, and worked as a cook in the local hospital. The hospital required its staff to receive the smallpox vaccination, but Maalin avoided it, he later said, on account of his fear of needles. So when Maalin volunteered to guide two smallpox-infected girls on their fifteen-minute car ride to an isolation camp, his immune system was unprepared for the viral onslaught.

Initially, local doctors misdiagnosed him with chicken pox and failed to isolate him. By this time the two girls had recovered so investigators mistakenly believed they had contained its spread. As a result, the popular Maalin had no fewer than ninety-one visitors stop by to wish him well. When doctors realized their mistake, the WHO launched a massive containment team throughout southern Somalia to track down everyone Maalin had come into contact with. They established police checkpoints, performed house-to-house searches, quarantined Maalin's entire hospital, and administered fifty thousand inoculations. And then they waited for a new case.

———

Four thousand years before Maalin's car ride, *variola* mutated into existence. The dominant strain began within a single person, in a single cell, in a single horrific moment. And through no fault of their own, this person may be responsible for more human death than anyone who has ever lived.

Who was he?

I'll call him Patient Zero, the code name doctors use for the originator of plagues. And I'll call him a him because the genetic clues within *variola* suggest he worked intimately with the domesticated camel near the Horn of Africa at a time and place where men and camels would often live for weeks in herding camps.

Patient Zero was born somewhere in modern-day Ethiopia or Eritrea nearly four thousand years ago, which is a few hundred years after the Egyptian pharaoh Khufu commissioned his

Great Pyramid one thousand miles to the north. Zero's culture had no writing, but direct accounts of the people of the Horn exist thanks to some of the oldest reliefs and hieroglyphs in ancient Egypt. The Egyptians called Zero's home the Land of Punt and Zero a Puntite, and both the Egyptians and the Puntites sailed the Red Sea along one of the world's first maritime trading routes.

If the Egyptian reliefs of the Puntite king Perahu are a faithful representation, Patient Zero may have had dark reddish skin with close-cropped hair worn beneath a skullcap. Like Perahu, he may have had a long goatee flipped forward. At his side, he may have tucked a dagger into his white bifurcated kilt.

Zero would have made his home in a round hut, according to the Egyptian reliefs, which he raised on stilts with a ladder to his door—perhaps to protect himself from predators or maybe to provide an enclosed corral for his animals underneath.

He was most likely a pastoralist who cultivated crops and raised a few domesticated animals, including dogs, cattle, donkeys, and camels. Myrrh trees, the sap of which could be burned as incense and was one of the most valued products of ancient Egypt, shaded his house. Demand for myrrh, along with Punt's gold, pelts, exotic animals, and slave labor drew the Egyptians to make the long voyage down the Red Sea. The pharaoh Khufu's son himself enslaved a Puntite person.

Zero likely lived and worked with his camel, though he did not ride it. Instead, he used it for its milk. A nursing camel can produce up to five gallons per day, and some camel herders have survived exclusively on it for weeks. Zero used his camel in the same tradition as the initial domesticators—a herding culture that occupied the southern Arabian Peninsula. Richard Bulliet, author of *The Camel and the Wheel*, believes soon after the camel's domestication five thousand years ago the camel made its way across the Red Sea on the boats of incense traders.

For the ancestor virus of *variola*, the introduction of the camel provided a tremendous opportunity; for humanity, an incredible risk. Any close contact with an animal's bodily fluids is a chance for a virus to hop species. Fortunately, different species have different immune systems, and a virus attuned to one is likely ill suited to attack another. Usually, when a virus finds itself in a foreign species, it's like "a human on Mars without a space suit," writes virologist Nathan Wolfe: it withers and dies. Or it's simply identified and destroyed by the new immune system. Or, more dangerously for the host, it replicates but is not contagious.

To successfully switch hosts, a virus must acquire variations that allow it to multiply and transmit in a new environment. A virus can do this in a variety of ways, either through its own mutations or, in some cases, two different viruses can infect the same cell and combine to form a new one.

By now, virologists believe we have been so thoroughly exposed to the primary diseases infecting dogs, pigs, and the rest of the domesticated animals that any disease of theirs that could feasibly hop into humans already has. Today, the primary viral danger from domesticated animals comes from their serving as a middleman between humans and the viruses of wild animals.

However, in Zero's era, the recent domestication and introduction of the camel into a new environment put the human species at risk. The camel gave the local viruses a new immune system to attack, and domestication brought day-to-day contact between humans and camels, providing millions of opportunities for viral introduction.

The precise ancestor of the *variola* virus is unknown. Donald R. Hopkins, a leader on the WHO's smallpox eradication team, suspects a poxvirus that infected rodents because of a particular gene found within *variola* for encoding a mouse protein. This is likely a remnant of a previous life, he muses in his book *The Greatest Killer.* Recent genetic research appears to support the notion that the direct ancestor virus to *variola* likely infected a now-extinct African rodent, perhaps a gerbil of some kind, but that the camel's immune system played the role of maître d'. Igor Babkin, a geneticist at the Institute of Chemical Biology in Novosibirsk, Russia, and author of the paper "The Origin of the Variola Virus," tells me the introduction of camels provided a stepping-stone from the rodents to humans. Babkin believes the process of attacking the camel's immune system triggered an evolution in the virus. As the gerbil's poxvirus replicated within the camel, he believes it picked up some new piece of genetic material that turned what had been merely a viral nuisance into a monster.

No one knows how Zero caught the first divergent strain from the African rodent. It could have been from butchering and eating it—as Hopkins speculates—or, as Stanford professor of immunology Robert Siegel tells me, it could have been from something as innocuous as breathing in the wrong puff of dust.

Dried rodent droppings become aerosolized in dust and can drift into lungs, which is how the hantavirus pulmonary syndrome is contracted.

Whatever the case, once inside Zero, the novel environment should have been the virus's end. Except this time, thanks to bad luck and some new trick it picked up from camels, the new poxvirus not only survived within Zero but thrived.

Once inside Zero's throat, *variola*'s viral grenade would have latched on to cells in his mucous membranes and begun replicating. Under a microscope, a *variola*-infected cell undergoes a hideous transformation, morphing from a smooth ball into a spiked jack. The hollow spikes stretch out from the cell like spears, hunting for healthy cells to impale. Once they strike, the hollow passageways within the spikes serve as tunnels for viral particles to hide from attacking white blood cells. Once inside the next cell, the process begins anew.

At first, Zero's immune system would have suspected nothing. Virologists aren't sure where the virus hides for the first seven to ten days, but many suspect it replicates within the lymph nodes, lulling the immune system into passivity while the virus grows exponentially. After more than a week of unabated growth, the viral particles burst forth and ride the bloodstream to the nearest organs. At this stage Zero's immune system would mount its first defense, and he would feel the first symptoms.

It would have started as a fever and headache, followed quickly by a sore throat, which would mark the beginning of Zero's infectiousness. Lesions in his throat would sprout viral grenades, each teeming with particles that would ride his microscopic spittle into the air to create a ten-foot viral hot zone. Initially he might have confused his symptoms for the flu, but when the virus began to attack his skin and distinctive red spots appeared on his neck, face, and back, he may have suspected something far more serious. Soon the spots would have filled with pus. Then he would have begun to smell.

William Foege, an epidemiologist on the WHO eradication team, writes in his book, *House on Fire*, that twice he was able to identify a smallpox patient inside a house from as far away as the sidewalk based only on the smell. He describes the odor as "reminiscent of the smell of a dead animal." No one knows what causes it. Perhaps, he proposes, it's the rotting pustules.

The pain is agony. Patients with smallpox wear almost nothing as any cloth's touch on their skin can burst open a pustule. At this stage, writes Foege, most patients simply want to die.

Variola didn't intend to kill Zero. According to Foege, "The virus carried no ill will; it was simply responding to the drive to perpetuate itself." Nevertheless, a patient infected with *variola* is in a fight to the death. Either Zero's immune system destroyed it, or *variola* would have destroyed him.

Immunologists are unsure of the lethality of the virus in the initial spillover event. The thousands of years it spent replicating in humans have changed it dramatically, but if *variola* followed the typical pattern, Zero's infection may have been an even deadlier version than the modern one.

Typically, the more recent a viral spillover, the more lethal, because from a viral perspective the death of its host is suboptimal. A virus seeks to replicate and spread, and a dead host no longer mingles with potential new ones. As a result, viruses often decrease in lethality the more time they spend in a host as more infectious variants become the dominant strain. Thousands of years after the initial spillover, a variation of *variola* called *variola minor* evolved, which had a death rate of only 1 percent compared to *variola major*'s 30 percent.

Unfortunately for Zero, he is likely to have had a far more lethal case than is modernly known and more likely than not succumbed.

But *variola* lived on. At some point during his infection—or even after his death if someone handled Zero's body—the virus found a new host to infect. In many cases, this would have been a family member. By modern times, when *variola* had fine-tuned itself to humans, a caring family member stood a 50 percent chance of catching the disease, according to Foege. But most virologists believe in its early stages *variola* was both more lethal and less adept at leaping between human hosts.

As Hopkins writes in *The Greatest Killer* about the initial infections, "If the pattern of adaptation of other diseases from animals to man may be presumed to have been followed in this case . . . spread from one person to another would have been rare at first."

Virologists refer to a virus's transmission rate as its R_0 (pronounced "R naught"). R_0 measures the average number of individuals a sick patient infects. A disease with an R_0 of less than one will eventually die out, while a disease with an R_0 of greater than one will spread. The higher the R_0, the greater the risk of an epidemic. Modern-day *variola* has an R_0 of six, which is roughly the equivalent of the common cold. Initially, Henderson believes, *variola*'s R_0 is likely to have been significantly lower, but its persistence is proof it was higher than one. And as it spread within humans, it adapted to them. It improved its transmission rate by working against our immune systems like a hacker brute-forcing a password. As it replicated trillions of times, and made millions of mistakes, the virus would occasionally stumble upon an improved variation and increase its rate of infection. Through natural selection, the new version would have soon become the dominant one and the process would repeat. There is no intelligence behind its adaptations. The virus is a million monkeys in front of keyboards, and *variola* is their *Hamlet*.

If Zero had simply lived in a small village, as humans have done for most of our history, the disease would have burned through the population and died out after either killing or immunizing everyone. Smallpox, like all highly infectious diseases, requires a large population to perpetuate itself. Researchers estimate *variola* required populations of at least two hundred thousand people within fourteen days of travel to self-sustain. Scholars believe there's little chance of *variola* succeeding prior to the agricultural revolution because population densities of that magnitude are a relatively recent development. Prior to farming, diseases too horrible to contemplate likely streaked through small communities, ran out of new hosts, and died.

But *variola* happened to leap into our species at one of the first times in history when populations of those numbers lived within that proximity—the dawn of the global economy, when pastoralists created some of the first large-scale communities in the

Horn of Africa and when the Egyptians and Puntites opened one of the first maritime trading routes. The disease could have percolated within the Horn for years before riding up the Red Sea in the body of an enslaved Puntite or one of the Egyptian traders. Once it landed in the Nile River valley, where four thousand years ago the population was over a million, the virus would have been unstoppable.

The first confirmed cases of smallpox come from the pustules virologists have identified on three Egyptian mummies. Two were anonymous Egyptian officials mummified 3,598 years ago and 3,218 years ago, while the third and most famous case is that of the pharaoh Ramses V, who died 3,175 years ago. Hieroglyphs describe Ramses dying of an acute illness, and when Henderson inspected the mummy he discovered smallpox's indicative pustules covering his face, neck, and shoulders. Once in Egypt, the disease spread throughout the world.

Approximately 3,400 years ago the Hittites from Turkey fought the Egyptian army, and their cuneiform tablets describe a disease brought on by their Egyptian captives. Egyptian traders along the incense trade route appear to have taken *variola* with them to India, where the ancient medical treatises *Charaka Samhita* and *Sushruta Samhita* describe smallpox in detail. *Variola* made it to China by 2,500 years ago, just before the building of the Great Wall, and in the sixteenth century, Spanish explorers introduced it to the New World, where its effects were particularly catastrophic. In outbreaks, immunized people reduce the R_0 of a virus and therefore decrease the disease's exponential growth. Because no one in the New World had survived an infection and become immunized, the initial outbreak was apocalyptic.

The fight against *variola* stretches back into antiquity. It was long understood that no one ever caught smallpox twice, but the Chinese were the first to use this weakness against the virus. According to the British biochemist and medical historian Joseph Needham, the first time that anyone anywhere in the world performed an inoculation—that is, intentionally infected a patient with a weakened form of a virus in order to confer immunity—occurred in China approximately one thousand years ago.

The first mention of inoculation is found in a fourteenth-century Chinese medical treatise:

"We do not know now the names of the inoculators, but they got it from an eccentric and extraordinary man who had himself derived it from the alchemical adepts. Since then it has spread widely all over the country."

This "extraordinary man" did nothing less than invent immunology, which remains humankind's greatest defense against the virus.

Variola's critical weakness is not only that a person cannot be infected twice but that dried pustules of a previously infected patient are naturally attenuated—meaning many of the viral particles have died off. Because a virus grows exponentially, fewer viral particles in the initial infection reduces the severity of the disease. The difference in mortality is dramatic. An inoculated patient stood a 1 to 2 percent chance of death compared to 30 percent for patients who were naturally infected.

By the seventeenth century, inoculation in China and India

was widespread. Doctors in China published manuals on the procedure and a century later knowledge of the practice made its way into Europe.

Inoculations, however, were far from perfect. If the doctor doesn't use properly dried scabs, for instance, they will infect their patient with full-strength smallpox. Even worse, an inoculated patient is as contagious and deadly as a traditionally infected one while their immune system fights off the disease. In many cases, inoculations sparked outbreaks.

The beginning of the end for *variola* occurred on May 14, 1796, when Dr. Edward Jenner tested a local inoculator's report of his inoculation's ineffectiveness on a few local farmers. These farmers swore they had never contracted smallpox but had recently come down with the bovine version of the poxvirus called cowpox. Cowpox infects humans, and the result isn't pleasant, but it doesn't spread human-to-human and is rarely lethal.

The inoculator's report sparked the beginnings of an idea in Jenner, and to test his theory he inoculated an eight-year-old boy named James Phipps with cowpox. Like in an inoculation of *variola*, Jenner attenuated the pustules of cowpox. The result was that Phipps didn't contract the full-blown, uncomfortable cowpox, but instead only a single, tiny pustule formed on his arm. And yet remarkably, when Jenner then administered a traditional *variola* inoculation, Phipps had no reaction at all.

Unlike *variola* inoculations, cowpox inoculations had a mortality rate of near zero, didn't cause subsequent outbreaks of smallpox, and only resulted in a single pustule rather than a week in the infirmary. Doctors later discovered an even better poxvirus for *variola* inoculation—called *Vaccinia virus*—and termed the procedure a vaccination.

Despite the vaccine, *variola* persisted. In the twentieth century alone it may have killed as many as 500 million people before an eighteen-year global hunt by the WHO isolated the virus in the body of Ali Maow Maalin. Luckily, Maalin's immune system gradually gained the upper hand against the virus. Even more remarkably, none of the ninety-one different people Maalin came in contact with came down with the disease.

So on November 1, 1977, the last cell of *variola* virus, scourge to humanity for nearly four thousand years, died at the hands of Maalin's immune system.

Addendum: At least two vials of the virus live on: one in a biolab in Atlanta, Georgia, and the other in Novosibirsk, Russia. No nation has acknowledged developing *variola* as a weapon, but in 1971 a Soviet field test of a smallpox bioweapon drifted onto a fishing vessel in Aralsk, killing three. And while Maalin was the last to contract the disease naturally, in 1978 at the University of Birmingham, England, *variola* escaped through a lab's ventilation system to an upstairs office, where it infected and killed a medical photographer named Janet Parker. Because widespread vaccinations for smallpox ceased in 1973, most of the world's population is once again vulnerable to the virus.

16

Who Told the First Joke We Know?

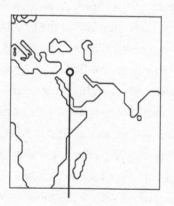

If the time of our species on earth
were a day, **this happened 19 minutes
before midnight** (4,000 years ago).

4,000 years ago

First known joke

On a November afternoon in 1872, in a corner of the British
Museum, the Assyriologist George Smith was hunched over an
ancient cuneiform tablet, slowly deciphering its symbols. The

work was painstaking. Cuneiform has hundreds of symbols and no punctuation, and more than two thousand years of decay had worn the etchings down to almost nothing. Still, he read with a focused anticipation. The damaged tablet had hinted at a secret, and he could barely wait while the museum's restorer rendered it legible. Finally, as Smith's finger fell upon the final symbol, he gave a shout. He leapt from his chair, raced about the room, and, according to legend, began removing his clothes. After ten years of searching, Smith had found his confirmation of the Old Testament's greatest tale.

Smith had an unlikely education for someone regarded as the era's leading Assyriologist. At fourteen he dropped out of school to begin work as a banknote engraver at the Bradbury & Evans publishing house, where he would have likely stayed had his office not been walking distance from the British Museum. Smith began to spend his lunch hour there, and his eye for etched detail drew him to the cuneiform tablets recovered from the ancient cities of Mesopotamia. Over lunches he taught himself to read the script well enough that the museum hired him away from his publishing job, and he quickly became one of the world's best Assyriologists.

In 1867, he discovered within the tablets a contemporary account of a solar eclipse and was able to identify it as the one astronomers had calculated to have occurred over Mesopotamia on June 15, 763 B.C. It's one of the oldest specific days ever recorded.

But as he continued to translate the ancient tablets, another obsession took hold. Smith began to search for firsthand confirmation of events described in the Old Testament—particularly Noah's flood. Such a catastrophic occurrence, he reasoned, would surely have been documented. Twenty years after he began his search, Smith found what he had been looking for. On a tablet discovered by the Ottoman archaeologist Hormuzd Rassam and written long before the Book of Genesis, Smith read of a "great flood" and of a man who saved the world's animals by building a giant boat—which is when Smith was compelled to rid himself of his clothing.

The flood story is part of *The Epic of Gilgamesh*, widely considered the oldest novel ever written. An anonymous Assyrian inscribed the flood story no later than 2,700 years ago—though there's no telling how long it existed in oral form. The story is fiction, but its similarity to Noah's flood is not a coincidence. Most scholars believe it inspired the Bible's version.

The Epic of Gilgamesh represents the final evolution of writing. For its first thousand years, writing in Mesopotamia had been the exclusive domain of accountants, serving largely to record debts and enforce taxes. The entire medium was the Mesopotamian equivalent of credit scores and tax forms, which raises for scholars the question: How did such a rigid accounting system morph into one that was endlessly flexible and able to record the world's best-known story 2,700 years ago?

The answer might be a few scribal school jokes.

Jokes and humorous proverbs first appeared nearly four thousand years ago, when master scribes wrote ancient quips for their students to copy, the Assyriologist Jana Matuszak tells me, both as a way for them to learn the then-dead language of Sumerian, and to impart morals like ancient versions of Aesop's fables. They began as simple sentences but evolved into complex tales of morality. And while it's nearly impossible to know which joke or proverb is the oldest because dating the cuneiform tablets can't be done with that level of precision, when I posed the question to Yale University professor of Assyriology Benjamin Foster, he nominated this four-thousand-year-old one-liner, which he believed to be as good a candidate as any for the world's oldest joke:

> When the lion came to the sheepfold, the dog put on his
> best leash.

It's profoundly unfunny. But since good jokes have a shelf life that is sometimes measured in days, this isn't a surprise. Whether it's funny today, however, isn't the point. The Akkadian creator received a chuckle for their prose, or at least they hoped to, according to Foster.

Thanks to this joke and others like it, humanity's most creative invention turned from a boring accountant's medium into something else entirely. Beginning with these early proverbs, writing became a way to communicate stories, information, and everything else the spoken word could conceivably conjure. Unfunny jokes marked the beginning of the medium's revolution from tax returns to comedy scripts and epic tales of ancient floods.

So who was this jokester who just may have saved written communication?

Let's call him Will, after one of the medium's greatest, Mr. Shakespeare, and a him because scholars note that the earliest proverbs are predominantly written from the male perspective, although female scribes certainly existed.

Will was born nearly four thousand years ago in the ancient city of Nippur, located in what is now central Iraq, roughly one hundred miles south of Baghdad. Nippur was founded some seven thousand years ago along the marshy banks of the Euphrates River and was one of the longest continuously occupied cities in the world before it fell into decay sometime in the first millennium. By the time of Will's birth, the river had shifted, the once frequently flooded plains had largely dried, and the citizens of Nippur had ceased using reed for their houses and had transitioned to the Mesopotamian staple of mud brick.

Nippur was a religious capital with a central temple. Most of the forty thousand Sumerian clay tablets archaeologists have recovered from the city deal with the temple's administration, the salaries of officials, the taxation of the citizens, and the accounting of grain. But a few clearly originate from Nippur's scribal schools, where temple administrators taught young students Sumerian. Will was one of these instructors, which means he occupied a relatively lofty position in Akkadian society and might have come from a family of some means. Perhaps he was the son of a wealthy merchant.

Akkadians were polytheists. They believed different gods controlled different and minute aspects of their daily life. If they had debts, they prayed to Utu to clear them. If they sought revenge, they would ask Ninurta for assistance. And if they bought a service from a shopkeeper, the shopkeeper might swear to Enki that the work would be done quickly. But Will himself may not have been terribly pious, according to Bendt Alster, who in his book *Wisdom of Ancient Sumer* wrote that the first proverbs display "a completely secular attitude toward social behavior." The consensus among scholars is that Sumerian scribes like Will were the Akkadian versions of a "Sunday Christian," men and women of negotiable religious conviction.

As a young man Will would have attended scribal school himself, an experience that appears to bear some remarkable similarities to the modern version of primary education. Children began at a similar age, had homework, took tests, and learned some of the same lessons taught in some of the same ways. Grammar would have been simpler, as cuneiform has no punctuation—though learning it would have been tedious nonetheless. Sumerian was by then a dead language, which according to Alster accounts for a few spectacular mistakes. One student trying to copy the symbols for "lame" apparently carved the symbol for "frog" instead, which is not a mistake a native speaker would make. Sumerian was the modern-day Latin of its time, existing only for the educated elite.

Scholars believe scribal schools taught far more than bookkeeping, however. Will also learned lessons on etiquette and morality and how a good Mesopotamian man or woman ought to behave. He did this largely through ancient oral proverbs intended to ridicule an archetypal fool and his vices or stupidity and thus teach better behavior. These old jokes and proverbs are part of what Matuszak tells me was "an art of insulting," which developed in ancient Mesopotamia and existed orally for centuries before Will inscribed the phrases on tablets, using the same symbols but with new purpose and meaning. In time, of course, the new idea became common among the scribes, and many examples have since been recovered, such as the following Akkadian slams translated and passed on by Matuszak:

... intelligence of a monkey!

The pigsty is her house. The oven is her reception room.

Your husband has no clothes to wear; you yourself are wearing rags. / Your butt sticks out from them!

Scribal school teachers told these humorous put-downs to elicit a chuckle, but also to impart lessons upon their students. "As becomes clear," Matuszak wrote me, "'humor' in Sumerian didactic literature isn't about telling jokes or funny anecdotes." The main aim is to educate students by presenting a protagonist who mocks and ridicules a fool, as in this example:

May my field be small, so I can go home (soon)!

A modern interpretation of this "joke" might be: "I hope I don't make too much income. That way I won't have to pay so much income tax!"

Will would have likely learned these proverbs and insults in oral form while as a student. And he must have learned them well, because he eventually advanced to scribal instructor, allowing him not just to follow lessons but to create them. His lesson, and innovation, was to put some of these insults and jokes into writing—perhaps to give his students a more interesting writing assignment than the endless documentation of barley shipments and business banalities they were used to. And the innovative joke he chose may well have been: When the lion came to the sheepfold, the dog put on his best leash.

Here's why Will thought that was funny: The dog is saying: "I used to be a guard dog, but now that I see this lion here, I'm switching back to a household pet." The humor is lost in the translation, but the bones of the joke are remarkably timeless. A modern retelling might include a lifeguard opting to take their lunch break as they see the fin cut through the water, or nearly the entire plot of *Sgt. Bilko.*

The fact that the oldest joke in recorded history shares so much in common with modern forms of humor is not lost on psychologists. There appears to be a certain bedrock formula to humor written somewhere in the *Homo sapiens* brain, which has led many scholars, philosophers, comedians, and writers to pursue a universal explanation of laughter and humor.

The Roman polymath Pliny the Elder believed laughter resulted from a stimulation of a diaphragm that stretched just under the skin, which he thought explained why people laugh when tickled under the thin-skinned armpits and—according to Pliny—why Roman gladiators died laughing when stabbed through the gut. More recently, psychologists have defined laughter broadly as the result of sudden "pleasant mental shifts." As a practical matter this is only slightly more explanatory than Pliny's diaphragm theory, so the methods jokes use to inspire these pleasant mental shifts are generally broken into three theories.

The guard dog joke relies upon what scholars call the incongruity theory of humor. A comedian sets your brain up for a zig—and then zags. For example, "A priest, a rabbi and a . . . duck walk into a bar, etc." A trio of MIT researchers led by Matthew Hurley suggest the pleasure we get from these incongruities is the result of a small dopamine reward our brain gives us for finding an error in our assumptions. The incongruity theory explains why you might pause after you read a simple joke like "A termite walks into a bar and asks, 'Is the bartender here?'" and find it a little funny.

The second theory of humor is what Aristotle described as the superiority theory.

The superiority theory describes the humor we experience when we suddenly feel better about ourselves. Typically, this comes at the expense of another and generally involves someone else's pain. Mel Brooks channeled Aristotle when he said, "Tragedy is when I cut my finger; comedy is when you fall into an open sewer and die." Expressed mathematically, the superiority theory reads: Comedy = Pain + Distance.

Finally, and most fundamentally, there is the relief theory. The American philosopher John Dewey defined the relief theory as "the sudden relaxation of strain." This theory explains why we find humor in a scary or unsettling event that suddenly transitions to a safe one, for example, a jack-in-the-box, a game of chase, and tickling. The relief theory works across species. Chimps enjoy chasing each other as much as any human toddler, which suggests the laugh that occurs at the sudden transition from terror to safety likely originates deep in antiquity.

Because humor often involves flipping one's expectations, and what one expects varies dramatically depending on time and culture, jokes reveal social structures. Who forms the butt of jokes, for example? Is it lawyers, barbarians, squires, or accountants? What about ethnicities, genders, or physical appearance? Thanks to the superiority theory, humor often kicks down at a society's most vulnerable. In cuneiform humor, enslaved people and women received a disproportionate amount of derision. Jokes like this one passed on by Matuszak, which is one of the oldest archaeologists have ever uncovered, hint at a misogynistic attitude in Akkadia:

> Her (ever so) pure womb is "finished"—(it means) financial loss for her house.

The phrase "ever so pure," according to Matuszak, is one of the oldest uses of written sarcasm. These old jokes reveal still more about Akkadian social structure. According to Foster, low-ranking officials and businessmen were frequent targets, while royalty were not. Those who displayed cowardice, ambition, bad manners, lust, and conceit were also fair game. Yet, as Foster tells me, "It's hard to identify downright obscene jokes in cuneiform when so little of the ancient standards of decency are known."

Violations of decency, including cursing, are a venerated tool of comics both ancient and modern. Cursing itself—the intentional violation of a cultural taboo—has likely always existed. Unfortunately, it's difficult to say what Will hollered after dropping a stone on his foot thanks to the ephemeral nature of taboos. What one culture finds incredibly offensive, the next might find milquetoast. Many of modern English's swear words, for example, are related to bodily functions and sex. But as recently as medieval England, privacy was so rare that the English performed many bathroom functions in the open and bathroom words lacked the cultural shock a swear word requires. Medieval kindergarten teachers might have mixed words like "piss" in with their daily lessons. Instead, when medieval Englishmen intended to shock they cursed "god's bones" or "god's blood." Would Will's obscene shouts have been of a religious nature, or ethnic, or bathroom related, or something else entirely? Scholars aren't sure.

But as much as punch lines change, the basic structure of humor and stories has varied little. In some cases, it hasn't at all. Take this four-thousand-year-old Akkadian story that Foster proposes as one of the oldest humorous tales ever written:

Nine wolves were holding ten sheep. There was one extra, so they did not know how to divide their shares. A fox having come upon them, said: "Let me divide your shares for you. You are nine, take one. I am one, let me take nine. That is the share I prefer."

The fox in this joke fits a character archetype anthropologists call the trickster, one who breaks society's rules not as a vandal but for their own personal gain. The trickster litters the folk stories of virtually every culture ever studied. In Native American tribes it's a role often played by a coyote, in African cultures by a rabbit, and in Hollywood by Jim Carrey or Bugs Bunny. It's a stock character that the *H. sapiens* brain seems to find timelessly funny and compelling.

The endurance of humor, profanity, and the trickster archetype suggest every *H. sapiens* culture has employed them, long before Will ever picked up a stylus. But he was the first to commit these jokes to a tablet, and in doing so these low-brow jokes and insults became some of the first non-accounting-related remarks ever written. Will's process of teaching cuneiform—perhaps in the interest of making it easier to learn—changed the medium itself. For centuries after Will, these written witticisms stayed within scribal schools. But as an increasing number of scribal teachers began writing these ancient Sumerian proverbs and didactic literatures, they developed a media of basic pictographs into one that could replicate every sound of language.

At first the written passages were simple one-liners. But as writing advanced, jokes and proverbs eventually became stories, myths, epic tales, songs, biographies, and medical treatises. With the small but not insignificant contribution of a creative Akkadian scribal schoolmaster like Will, written history itself was launched.

Unfortunately, Will's tomb is probably lost forever. Thanks to his contribution, though, the hard evidence of his middling sense of humor is not and never will be.

Who Discovered Hawaii?

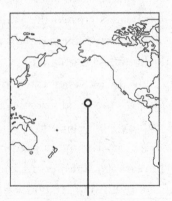

If the time of our species on earth were a day, **this happened 5 minutes before midnight** (1,000 years ago).

1,000 years ago

First foot on Hawaii

From one perspective, the out-of-Africa history of modern *Homo sapiens* is one of global expansion and exploration. It began approximately fifty-five thousand years ago, when modern *H. sapiens* burst forth from Africa across the Sinai Peninsula and into the Middle East and spread in a diaspora across Europe and Asia.

Some went north, migrating through the European continent and, based on the evidence of a fossilized human jawbone, made it to England (using a land bridge) by 41,000 years ago. Another group went south. They passed through India, through Southeast Asia, and across the now-submerged land bridges between the Indonesian islands of Sumatra, Java, and Bali, until a deep-sea trench east of Borneo called the Wallace Line halted their progress. Even with seas nearly three hundred feet lower than today, passage across the Wallace Line required boats. Yet, judging by *H. sapiens* teeth on the island of Flores, archaeologists believe these *H. sapiens* boaters crossed the Wallace Line at least 46,000 years ago, and by forty thousand years ago were through Australia and to the bottom of Tasmania.

Another group moved east. They traveled across Asia and populated Japan by forty thousand years ago, moved north of the 60th parallel—above which no previous hominin had ever survived—into western Siberia and crossed onto the Beringia landmass by 32,000 years ago. The massive ice sheets of Canada blocked their passage south until 16,000 years ago, when the earth warmed and the ice melted. Once beyond Canada, the first Americans reached the bottom of South America within a millennium. By four thousand years ago, humans landed upon the eastern edge of arctic Canada and Greenland.

By three thousand years ago, modern humans occupied nearly the entire habitable planet. Only the lush, remote islands of the South Pacific remained vacant.

After the early boaters crossed the Wallace Line nearly forty-five thousand years ago, they continued east. At first, the islands were so close that a person standing on the beach could see the next one. These tightly packed isles provided what professor of archaeology at the University of Auckland Geoffrey Irwin calls a "voyaging nursery," where boaters developed their crafts and navigational techniques in the relative safety of the vast archipelagos.

Using basic sailboats, these early adventurers hopped islands until they reached the Solomon Isles, just east of Papua New Guinea. Past the Solomons, however, the islands of the South Pacific become increasingly sparse. The next island is Vanuatu, which is separated from the Solomons by a daunting three hundred miles of open ocean.

The distance halted all human progress into the eastern Pacific for thousands of years, until perhaps human history's most voracious explorers—archaeologists call them the Lapita culture—burst south from Taiwan approximately three thousand years ago. Little evidence remains of the Lapita's boats, but scholars speculate they may have invented the outrigger canoe, which provided increased stabilization and improved their performance into the wind. Using their outriggers, the Lapita moved through the Solomons, discovered Vanuatu, and then sailed beyond. Within four hundred years they discovered Fiji, West Polynesia, Tonga, and Samoa. They hopped islands so quickly archaeologists believe the only possible explanation for their movements is the desire to explore. According to Irwin, "They went . . . because they wanted to see what was over the horizon."

Eventually, the vastness of the remote Pacific stalled even the Lapita. Beyond Samoa, scattered and distant, are the islands of New Zealand, the Cook Islands, Tahiti, the Marquesas Islands, Easter Island, Hawaii, and finally the coast of South America. The distances are measured in the thousands of miles, and the voyages required large boats, months at sea, and expert navigators. Yet by 1,200 years ago the Polynesian culture overcame these incredible stretches of open ocean.

Polynesians sailing to the south discovered the Cook Islands and then New Zealand. Others went east, first to Tahiti, then the Marquesas, Easter Island, and finally, proven by the presence of sweet potatoes in New Zealand, they reached the Chilean coast of South America. Sweet potatoes are native only to the Americas, and their presence in Polynesian farms in New Zealand proves at least a group of Polynesian explorers not only visited South America but maintained trading relationships across nearly six thousand miles of open ocean.

But of all their navigational accomplishments, the Polynesian discovery of Hawaii stands as perhaps the greatest feat of exploration in human history.

Hawaii is the world's most isolated archipelago. It's a geologic hiccup in a vast stretch of nothing. It's so far from anywhere that many European explorers, upon discovering it occupied, believed its discoverers must have been shipwrecked. But computer simulations, archaeological discoveries, and re-creations of the voyage have removed all doubt: The discovery of Hawaii was no accident. Not only is there no evidence for that theory, but according to the archaeologist Scott Fitzpatrick, simulations of the

Pacific's trade winds and currents suggest that an undirected voyage—regardless of its land of origin—would essentially have zero chance of landing on Hawaii. Early Western theories of Hawaii's accidental discovery had it backward. Hawaii's isolation isn't the reason it must have been found accidentally, it's the reason why it couldn't have been.

According to Fitzpatrick, Hawaii could have only been discovered on a direct search, using a fast boat departing in the early fall months, most likely from the Marquesas.

In other words, the colonizers of the South Pacific were epic seafaring explorers, and whoever discovered Hawaii may have been the greatest of them all.

Who was it?

I'll call him Captain Kirk, after the great *Star Trek* explorer of new worlds.

Kirk was born nearly 1,000 years ago on the Marquesas Islands, which are themselves an isolated island chain 2,300 miles southeast of Hawaii. Very few contemporaneous accounts exist of Marquesan culture. Eastern Polynesians of Kirk's era didn't use a writing system, and when European missionaries arrived in the eighteenth century their diseases—primarily smallpox—killed 98 percent of the local inhabitants. As a result, much of what anthropologists know of this ancient navigating culture has been pieced together from archaeological evidence, diary entries by some of the first Europeans to visit, neighboring Polynesian cultures, and the topography of the islands themselves.

The Marquesan Islanders were pastoralists and fishermen, not hunter-gatherers. Kirk raised domesticated pigs and chickens, cultivated breadfruit, taro, and yams, and fished and gathered shellfish along the plentiful reefs of the Marquesas. His clothing was simple. The islands are tropical and warm year-round, so Kirk may have worn as little as a bark-cloth loin wrap, but he decorated almost every other aspect of his appearance.

In 1774 Charles Clerke, a sailor aboard James Cook's famous expedition to the South Pacific, wrote that the Marquesans were "tattow'd from head to food in the prettyest manner that can be conceiv'd." Marquesans took their tattoos seriously, as expressions of their culture and religious beliefs, and received them throughout their lives. Tattoo artists were full-time professionals who punched charcoal ink into skin with bone needles and a hammer stick. Their clients paid them in food, goods, and services, and their artistry was both expensive and painful. The Marquesan word for tattoo was "ta-tu," and when the European sailors returned home, they adopted both the art and the word.

Kirk would almost have certainly been heavily tattooed with pierced ears hung with ivory. On special occasions he wore flowers, a headdress of rooster tails, or a crown made of porpoise teeth so elaborate it took weeks to make. He decorated everything he used, since the Marquesan Islanders believed their art wasn't just aesthetic, it was a means to communicate with their many gods.

He may have also been a musician. Marquesans played nose flutes, and European explorers' accounts suggest the men often serenaded women from outside their window like lovesick teenagers, using nose flutes instead of boom boxes and guitars.

Kirk was, no doubt, a warrior as well.

The Marquesas are high, dry islands with mountains that rise to form steep-walled, protected valleys. High islands are notoriously violent. The walls inhibit communication among groups, and the isolated communities can easily devolve into incessant revenge killings. Their battles were far smaller in scope and casualties compared to the organized, state-sponsored warfare of twentieth-century Europe. But anthropological observations have found the ceaseless violence often resulted in per-capita homicide rates five times that of even twentieth-century Europe.

As a consequence of the violence, Kirk may have practiced human sacrifice and perhaps cannibalism. Early anthropologists' reports suggest both occurred on the Marquesas, and, according to the political scientist James Payne, human sacrifice is a predictable result of endemic violence. Despite stereotypes, its practice was far from unique to the South Pacific or their religions. Virtually every major religion has practiced ritual killing at one time or another, according to Payne, because the practice has little to do with the teachings of a particular religion and is instead a symptom of chronic violence. According to Payne, frequent violence has the dual effect of cheapening the value of life and, because of its ubiquity, convincing citizens their gods must favor it. If life is cheap, the theory holds, paying one for a god's favor seems affordable.

More than a warrior, farmer, fisherman, or musician, though, Kirk was a sailor.

Eastern Polynesians had largely abandoned long-distance voyages by the time Captain Cook arrived in Hawaii in 1779, so there are few sources to tell us how Kirk learned navigation.

But from the accounts of Tupaia, a Tahitian navigator who rode aboard Cook's *Endeavour*, and the stories and experiences of the traditional navigator Mau Piailug, who was born in 1932 on the Micronesian island of Satawal, scholars have been able to piece together a likely story.

Kirk may have learned to sail from his father or grandfather, as did Piailug, who began his education when his grandfather taught him the ports, winds, and currents of the South Pacific. Kirk would have learned to navigate using the stars, and judging by Tupaia's firsthand accounts, Kirk's education was more akin to a PhD in astronomy than learning the constellations by the campfire.

While aboard Captain Cook's HMS *Endeavour*, Tupaia informed Cook of the bearings, sailing times, dangerous reefs, harbors, and chiefs on all of the Society Islands, Austral Islands, Cook Islands, Samoa, Tonga, Tokelau, and Fiji. Contemporary accounts differ, but Tupaia either drew or dictated a map from memory that covered 10 million square miles of the South Pacific and its 130 islands. Joseph Marra, a midshipman on the *Endeavour*, described Tupaia as "a man of real genius," which is high praise from a group not known for their open-mindedness toward foreign cultures.

Because there was no writing in Polynesian culture, nor any maps, Kirk would have had to memorize a prodigious amount of information. He had to know the stars and their arcs. He had to know in what seasons they appeared, when they disappeared, and which stars made their zeniths above which islands. Of the Polynesian navigational ability, Cook wrote:

The clever ones among them will tell in what part of the heavens they are to be seen in any month when they are above their horizon. They know also the time of their annual appearing and disappearing to a great nicety, far greater than would be easily believed by a European astronomer.

Elder navigators would have tested Kirk on the water. He would have been sent on solo navigations where losing one's way would have meant certain death. He would have sailed frequently to the neighboring islands in the Marquesas chain, and likely even traversed the 850-mile crossing to Tahiti. His boat is sometimes called a canoe, but I think something is lost in the English translation, and it has led to a misunderstanding of Polynesian sailing capabilities that may have allowed the undirected float theory to propagate. A "canoe" in English is something you use for day outings on a flat lake, but that's a far cry from Kirk's craft.

Kirk's "canoe," based on uncovered ancient hulls and oral histories, was an eighty-foot double-hulled catamaran complete with a stove, at least two sails, and a shelter on deck. It was a formidable craft I'll call the *Enterprise*.

Kirk built the *Enterprise*'s hulls from the massive tamanu tree and deepened the planks to increase their carrying capacity. He wove coconut husks for rigging and made giant sails out of matted palm tree leaves. Scholars believe constructing these enormous boats required so many resources, island exploration was not merely a side pursuit of the Polynesian culture but an organizing principle.

The finished *Enterprise* could hold forty people and enough food and water stored in coconut shells to last for months. The water alone would have weighed more than ten thousand pounds. *Enterprise* would have been capable of carrying an island occupation starter kit, including breeding pairs of pigs, chickens, sweet potatoes, yams, breadfruit, and seeds, but on Kirk's initial voyages of exploration failure would have been so frequent he would likely have taken only the crew and supplies he needed to complete a search.

By middle age, Kirk was an experienced open-ocean navigator. He may have already sailed both east—to Easter Island—and west—to Tahiti and the Cook Islands. But on the voyage on which he discovered Hawaii, he decided to sail north. The Polynesians knew of no land in that direction, leaving his inspiration a mystery. Perhaps, as the archeologist Patrick Kirch proposes in his book *A Shark Going Inland Is My Chief,* the flight of a bird—the Pacific golden plover—caught his eye.

The Pacific golden plover is a small brown-and-yellow-checkered bird that forages along beaches and tidal flats. For six months it makes its home in the Marquesas before it flies north in April. Six months later, the bird returns. When Captain Cook observed the plover flying above his ship in 1778, he noted in his journal, "Does this not indicate that there must be land to the north where these birds retired in the proper season to breed?" Kirk might have asked himself the same question nearly a millennium before.

Yet, in this observation he was lucky. The plover's remarkable migration does not stop in Hawaii, but continues unabated to the coast of Alaska. Still, its flight north could have inspired hope of a distant land.

According to simulations of Kirk's voyage performed by Fitzpatrick's weather models, Kirk would have stood the best chance of discovering Hawaii if he departed in November, when the trade winds are favorable. If he left any time other than the fall the models give him no chance of sailing the required distance before he would have had to return home.

As Kirk sailed, he navigated by tracking his latitude and longitude. Calculating his latitude was relatively straightforward: Kirk simply measured the apex of the night stars or the noon sun above the horizon. Estimating his longitude was another matter entirely. To do so, he likely used a method of navigation sailors grimly refer to as dead reckoning, which simply means he determined his location by the direction of the winds, swells, the flight path of migrating birds, the positions of the stars, the arc of the sun, and by estimating his speed through the water while taking into account the currents. Necessarily, this rules out sleep. Modern dead reckoners stay at the tiller for twenty-two hours a day, and rely on a trusted mate for the remainder, in order to track their speed and the current.

As he sailed, Kirk would be on the careful lookout for indications of a nearby island. Near-shore birds, a refracting swell, sea turtles, floating debris, or a stack of clouds piling up on the horizon could have alerted him to the presence of land. Given typical conditions, computer simulations estimate Kirk may have averaged three knots on his journey, which means after twenty-four days at sea he may have spotted gulls, a change in swell angle, or clouds stacking up against Hawaii's nearly fourteen-thousand-foot Mauna Kea.

No one knows where Kirk made his landfall. Captain Cook chose Kauai's Waimea Bay on his first voyage, and the Big Island's Kealakekua Bay for his second, but wherever Kirk disembarked, he would have found an astonishing Eden. Conches the size of dinner plates littered the reefs, and large flightless birds made for easy meals. Yet he probably didn't stay for long. Most scholars believe Polynesians' initial voyages were exploratory, so Kirk likely refilled his supplies, confirmed it was a habitable island, and returned home.

Navigationally, his return would have been the more perilous. After a month of sailing he would have arrived at the proper latitude, but then would have faced a terrible choice. Was he east of the Marquesas or west?

Kirk's dilemma was referred to by sailors until the late-seventeenth century as the "problem of longitude," which refers to the lack of celestial clues the sky offers a sailor as to their east-to-west location, save for the imperceptible differences in the time of the sun's rise and set (modernly experienced as time differences or jet lag). If Kirk made a small error in the calculations of his speed or current, or if a storm blew him off course, he could easily have made the wrong choice and sailed toward certain death. His life depended upon every observation he had taken on his voyage to keep him from sailing into oblivion.

Seven hundred years after Kirk, the British commodore George Anson found himself in this exact predicament while captaining the *Centurion*. Storms below Tierra del Fuego had disoriented his navigation and waylaid his ship, and as his crew died of scurvy he searched for respite in the Juan Fernández Islands. He captained his ship to the proper latitude, but no longer sure of his longitude, he didn't know whether to sail east or west. He chose both, spending days zigzagging off the coast of a hostile Chile while eighty of his crew died before he found port.

When Captain Cook himself sought the Marquesas Islands, he relied upon the previous measurements of Captain Álvaro de Mendaña, who had been the first European to arrive there, in 1595. Mendaña had noted the correct latitude of the islands, but his longitude allowed for a 685-mile margin of error—a margin approximately the width of the state of Texas. Cook found the islands because he originated from a location he knew was east of the Marquesas, which meant he simply sailed to the proper latitude and then straight west until he sighted land—a technique known as sailing down the longitude.

Because Kirk was returning from the north, this would not have been an option. When he reached the proper latitude, he would not have had time to sail down a Texas-size error in his estimate. He would have had to know whether to sail east or whether to sail west.

European sailors weren't able to measure their longitude until 1761, when an English clockmaker named John Harrison constructed a watch that could keep stunningly accurate time for months. How Kirk did it is somewhat of a mystery. Some scholars speculate that he may have even had a way of determining his longitude by some unknown, long-lost technique developed over the thousands of years of island colonization. As Irwin writes in "Voyaging and Settlement," "It's possible . . . these great navigators were aware of the zenith stars of many islands, and could form an algorithm that allowed a sailor to sail west of their home island without fear."

Once Kirk returned to the Marquesas, it's likely he led a colonizing group back—perhaps in several boats, each carrying forty people along with the animals and seeds they would need to start a new colony. When they arrived back on Hawaii, the carbon record suggests they burned vast tracts of forest for their farms and hunted the island's large flightless birds to extinction within a generation. The virgin territory and fertile land sparked a population explosion. By the time Cook arrived nearly eight hundred years later, an estimated half a million people lived on the islands.

As for Kirk, he probably continued sailing. For centuries, the eastern Polynesian islands kept in contact with each other. Polynesian oral histories tell the stories of ancient navigators who sailed between Tahiti and Hawaii, and there are ancient stone tools on Tahiti made from Hawaiian rocks. But eventually, the tradition died off. Perhaps, as Kirch proposes, once the Hawaiian population reached a certain threshold, they refocused their efforts on interisland sailing. Dozens of boats greeted Cook as he sailed into Kauai, but none of them were on the scale of the *Enterprise*. Yet even though the ancient catamarans were gone, Cook himself had no doubts regarding the Polynesians' navigational abilities. He always believed their colonization had been intentional.

In 2014, archaeologists found a small piece of skull in a cave in Israel dated 55,000 years old. This small piece of bone represents the beginning of modern *Homo sapiens* exploration of the habitable globe.

Fifty-four thousand years later, Kirk's discovery of Hawaii represented the end.

Acknowledgments

I could not have contemplated or even approached this idea without the extraordinary generosity of so many from the academic community who provided me with their patience, research, and time. I relied upon far too many people to list here, but there are a few who I leaned on more than most. It goes without saying that every mistake I made in the previous pages is completely my own, but rest assured I would have made far more were it not for the considerable assistance of those listed below.

Thanks to Bill Durham for his early inspiration, introductions, and guidance when this book was only an idea. Thanks to Cara Wall-Scheffler and Tate Paulette for their ideas, comments, and corrections. Thanks to Keith Devlin, John Rick, Seth Rasmussen, Richard Klein, and Robert Siegel—and his students!—for their time and patient explanations. Thanks to Patrick Hunt for his help in solving an unusually cold case. And thanks to Jana Matuszak for cracking a few of her very, very, old one-liners.

Thanks to the many family and friends who read drafts and offered ideas. Thanks to Kevin Plottner for his predictable aptitude in cartography and horology. Thanks to Alia Hanna Habib for her representation, guidance, and encouragement. Thanks to Shannon Kelly and the many members of the Penguin team who helped shepherd this book through every stage. And many thanks to my wonderful editor Meg Leder for so skillfully guiding these stories.

Sources and Further Reading

Who Invented Inventions?

Byrne, Richard. *The Manual Skills and Cognition That Lie Behind Hominid Tool Use.* Cambridge, UK: Cambridge University Press, 2004.

Currier, Richard. *Unbound: How Eight Technologies Made Us Human and Brought Our World to the Brink.* New York: Arcade Publishing, 2015.

DeSilva, Jeremy M. "A Shift Toward Birthing Relatively Large Infants Early in Human Evolution." *Proceedings of the National Academy of Sciences* (January 2011).

Falk, Dean. "Prelinguistic Evolution in Early Hominins: Whence Motherese?" *Behavioral and Brain Sciences* (August 2004).

Goodall, Jane. *The Chimpanzees of Gombe: Patterns of Behavior.* Cambridge, MA: Harvard University Press, 1986.

Henrich, Joe. *The Secret of Our Success: How Culture Is Driving Human Evolution, Domesticating Our Species, and Making Us Smarter.* Princeton, NJ: Princeton University Press, 2015.

Rosenberg, Karen, et al. "Did Australopithecines (or Early Homo) Sling?" *Behavioral and Brain Sciences* (August 2004).

Stringer, Chris. *Lone Survivors: How We Came to Be the Only Humans on Earth.* New York: Times Books, 2012.

Scheffler, Cara-Wall, et al. "Infant Carrying: The Role of Increased Locomotory Costs in Early Tool Development." *American Journal of Physical Anthropology* (June 2007).

Taylor, Timothy. *The Artificial Ape: How Technology Changed the Course of Human Evolution.* New York: St. Martin's Press, 2010.

Walter, Chip. *The Last Ape Standing: The Seven-Million-Year Story of How and Why We Survived.* New York: Bloomsbury, 2013.

Wong, Kate. "Why Humans Give Birth to Helpless Babies." *Scientific American,* August 28, 2012.

Who Discovered Fire?

Aeillo, Leslie, and Peter Wheeler. "The Expensive-Tissue Hypothesis." *Current Anthropology* (April 1995).

"Bonobo Builds a Fire and Toasts a Marshmallow." *Monkey Planet,* BBC One. 2014.

Bramble, Dennis, and Daniel Lieberman. "Endurance Running and the Evolution of Homo." *Nature* (November 2004).

Carmody, Rachel N., and Richard W. Wrangham. "The Energetic Significance of Cooking." *Journal of Human Evolution* (2009).

Hoberg, E. P., et al. "Out of Africa: Origins of the Taenia Tapeworms in Humans." *Proceedings of the Royal Society Biological Sciences* (April 2001).

McGee, Harold. *On Food and Cooking: The Science and Lore of the Kitchen.* New York: Scribner, 1984.

Plavcan, J. Michael. "Body Size, Size Variation, and Sexual Size Dimorphism in Early Homo." *Current Anthropology* (December 2012).

Pruetz, Jill D., and Nicole M. Herzog. "Savanna Chimpanzees at Fongoli, Senegal, Navigate a Fire Landscape." *Current Anthropology* (August 2017).

Raffaele, Paul. "Speaking Bonobo." *Smithsonian Magazine,* November 2006.

Salzberg, Hugh. *From Caveman to Chemist: Circumstances and Achievements.* Washington, DC: American Chemical Society, 1991.

Sorensen, A. C. et al. "Neanderthal Fire-Making Technology Inferred from Microwear Analysis." *Scientific Reports* (July 2018).

Theunissen, Bert. "Eugène Dubois and the Ape-Man from Java." Dordrecht, The Netherlands: Kluwer Academic Publishers Group, 1988.

Wade, Nicholas. *Before the Dawn: Recovering the Lost History of Our Ancestors.* New York: Penguin Press, 2006.

Wrangham, Richard. *Catching Fire: How Cooking Made Us Human.* New York: Basic Books, 2009.

Who Ate the First Oyster?

Caspari, Rachel, and Sang-Hee Lee. "Older Age Becomes Common Late in Human Evolution." *Proceedings of the National Academy of Sciences* (July 2004).

Frisch, Rose, and Janet McArthur. "Menstrual Cycles: Fatness as a Determinant of Minimum Weight for Height Necessary for Their Maintenance or Onset." *Science* (October 1974).

Kelly, Robert L. *The Lifeways of Hunter Gatherers: The Foraging Spectrum.* Cambridge, UK: Cambridge University Press, 2013.

Klein, Richard, with Blake Edgar. *The Dawn of Human Culture.* New York: John Wiley, 2002.

Lee, Richard Borshay. *The !Kung San: Men, Women, and Work in a Foraging Society.* Cambridge, UK: Cambridge University Press, 1979.

Marean, Curtis, et al. "Early Human Use of Marine Resources and Pigment in South Africa During the Middle Pleistocene." *Nature* (October 2007).

National Cancer Institute. "Age and Cancer Risk." Cancer.gov. April 2015.

Shea, John. "The Middle Paleolithic of the East Mediterranean Levant." *Journal of World Prehistory* (January 2003).

Trinkaus, Erik, et al. "Anatomical Evidence for the Antiquity of Human Footwear Use." *Journal of Archaeological Science* (August 2005).

Who Invented Clothing?

Brown, Donald E. *Human Universals.* New York: McGraw-Hill Humanities/Social Sciences/Languages, 1991.

deMenocal, Peter, and Chris Stringer. "Climate and the Peopling of the World." *Nature* (October 2016).

Gilligan, Ian. "The Prehistoric Development of Clothing: Archaeological Implications of a Thermal Model." *Journal of Archaeological Method and Theory* (January 2010).

Hayden, Brian. "Practical and Prestige Technologies: The Evolution of Material Systems." *Journal of Archaeological Method and Theory* (March 1998).

Hogenboom, Melissa. "We Did Not Invent Clothes Simply to Stay Warm." BBC Earth. September 19, 2016.

Kittler, Ralf, Manfred Kayser, and Mark Stoneking. "Molecular Evolution of Pediculus Humanus and the Origin of Clothing." *Current Biology* (August 2003).

Papagianni, Dimitra, and Michael A. Morse. *The Neanderthals Rediscovered: How Modern Science Is Rewriting Their Story.* New York: Thames & Hudson, 2013.

Pinker, Steven. *The Blank Slate: The Modern Denial of Human Nature.* New York: Penguin, 2003.

Rodgers, Alan, et al. "Genetic Variation at the MC1R Locus and the Time since Loss of Human Body Hair." *Current Anthropology* (2004).

Ruddiman, William. *Earth's Climate: Past and Future.* New York: W. H. Freeman, 2003.

St. Clair, Kassia. *The Golden Thread: How Fabric Changed History.* London: John Murray, 2018.

Toups, Melissa, et al. "Origin of Clothing Lice Indicates Early Clothing Use by Anatomically Modern Humans in Africa." *Molecular Biology and Evolution* (January 2011).

Wales, Nathan. "Modeling Neanderthal Clothing Using Ethnographic Analogues." *Journal of Human Evolution* (December 2012).

Zinsser, Hans. *Rats, Lice and History.* Boston, MA: Little, Brown and Company, 1934.

Who Shot the First Arrow?

Alexander, Gerianne, and Melissa Hines. "Sex Differences in Response to Children's Toys in Nonhuman Primates (*Cercopithecus aethiops sabaeus*)." *Evolution and Human Behavior* (November 2002): 464–79.

Backwell, Lucinda, et al. "The Antiquity of Bow-and-Arrow Technology: Evidence from Middle Stone Age Layers at Sibudu Cave." *Antiquity* (April 2018).

Churchill, Steven. "Weapon Technology, Prey Size Selection, and Hunting Methods in Modern Hunter-Gatherers: Implications for Hunting in the Palaeolithic and Mesolithic." *Archaeological Papers of the American Anthropological Association* (January 1993).

Churchill, Steven, et al. "Shanidar 3 Neandertal Rib Puncture Wound and Paleolithic Weaponry." *Journal of Human Evolution* (August 2009).

Farmer, Malcolm. "The Origins of Weapon Systems." *Current Anthropology* (December 1994).

Kennett, Douglas. "Sociopolitical Effects of Bow and Arrow Technology in Prehistoric Coastal California." *Evolutionary Anthropology Issues News and Reviews* (May 2013).

Kratschmer, Alexandra Regina, et al. "Bow-and-Arrow Technology: Mapping Human Cognition and Perhaps Language Evolution." *The Evolution of Language: The Proceedings of the 10th International Conference*. 2014.

Kroeber, Theodora. *Ishii in Two Worlds: A Biography of the Last Wild Indian in North America*. Berkeley and Los Angeles: University of California Press, 1961.

Lombard, Marlize. "Indications of Bow and Stone-Tipped Arrow Use 64,000 Years Ago in KwaZulu-Natal, South Africa." *Antiquity* (September 2010).

Lombard, Marlize, and Miriam Noël Haidle. "Thinking a Bow-and-Arrow Set: Cognitive Implications of Middle Stone Age Bow and Stone-Tipped Arrow Technology." *Cambridge Archaeological Journal* (June 2012).

Sisk, Matthew, and John Shea. "Experimental Use and Quantitative Performance Analysis of Triangular Flakes (Levallois Points) Used As Arrowheads." *Journal of Archaeological Science* (September 2009).

Wadley, Lyn, et al. "Traditional Glue, Adhesive and Poison Used for Composite Weapons by Ju/'hoan San in Nyae Nyae, Namibia. Implications for the Evolution of Hunting Equipment in Prehistory." *PLoS ONE* (October 2015).

Yu, Pei-lin. "From Atlatl to Bow and Arrow: Implicating Projectile Technology in Changing Systems of Hunter-Gatherer Mobility." In *Archaeology and Ethnoarchaeology of Mobility*, edited by Pei-lin Yu. Gainesville, FL: University of Florida Press, 2006.

Who Painted the World's First Masterpiece?

McAuliffe, Kathleen. "If Modern Humans Are So Smart, Why Are Our Brains Shrinking?" *Discover* (September 2010).

Cave of Forgotten Dreams. DVD. Directed by Werner Herzog. Orland Park, IL: MPI Media Group, 2011.

Chauvet, Jean-Marie, Eliette Brunel Deschamps, and Christian Hillaire. *Dawn of Art: The Chauvet Cave*. New York: Harry N. Adams, 1996.

Clottes, Jean. *Cave Art*. New York: Phaidon Press, 2010.

Conard, Nicholas. "New Flutes Document the Earliest Musical Tradition in Southwestern Germany." *Nature* (August 2009).

Curtis, Gregory. *The Cave Painters: Probing the Mysteries of the World's Prehistoric Artists*. New York: Random House, 2008.

Franses, P. H. "When Do Painters Make Their Best Work?" *Creativity Research Journal* (2013): 457–62.

Fu, Qiaomei, et al. "The Genetic History of Ice Age Europe." *Nature* (June 9, 2016).

Lotzof, Kerry. "Cheddar Man: Mesolithic Britain's Blue-Eyed Boy." Natural History Museum website, February 7, 2018, updated April 18, 2018. https://www.nhm.ac.uk/discover/cheddar-man-mesolithic-britain-blue-eyed-boy.html.

Robinson, John. "Return to the Chauvet Cave." Bradshaw Foundation website, 2001.

Théry-Parisot, Isabelle, et al. "Illuminating the Cave, Drawing in Black: Wood Charcoal Analysis at Chauvet-Pont d'Arc." *Antiquity* (April 2018).

Thurman, Judith. "First Impressions," *New Yorker,* June 23, 2008.

Who First Discovered the Americas?

Abbott, Alison. "Mexican Skeleton Gives Clue to American Ancestry." *Nature* (May 2014).

Bourgeon, Lauriane, et al. "Earliest Human Presence in North America Dated to the Last Glacial Maximum: New Radiocarbon Dates from Bluefish Caves, Canada." *PLoS One* (January 2017).

Cinq-Mars, Jacques, and Richard E. Morlan. "Bluefish Caves and Old Crow Basin: A New Rapport." In *Ice Age Peoples of North America. Environments, Origins, and Adaptations of the First Americans,* edited by Robson Bonnichsen and Karen L. Turnmire. College Station, TX: Texas A&M University Press, Center for the Study of the First Americans, 1999.

Erlandson, Jon, et al. "The Kelp Highway Hypothesis: Marine Ecology, the Coastal Migration Theory, and the Peopling of the Americas." *Journal of Island and Coastal Archaeology* (October 2007).

Fagundes, Nelson, et.al., "How Strong Was the Bottleneck Associated to the Peopling of the Americas? New Insights from Multilocus Sequence Data." *Genetics and Molecular Biology* 41 (2018): 206–14.

Goebel, Ted. "The Archaeology of Ushki Lake, Kamchatka, and the Pleistocene Peopling of the Americas." *Science* (2003).

Graf, Kelly E., Caroline V. Ketron, and Michael R. Waters. *Paleoamerican Odyssey.* College Station, TX: Texas A&M University Press, 2014.

Lesnek, Alia, et al. "Deglaciation of the Pacific Coastal Corridor Directly Preceded the Human Colonization of the Americas." *Science Advances* (May 2018).

Potter, Ben A. "A Terminal Pleistocene Child Cremation and Residential Structure from Eastern Beringia." *Science* (February 2011).

Pringle, Heather. "Welcome to Beringia." *Science* (February 28, 2014).

Ruhlen, Merritt. *The Origin of Language: Tracing the Evolution of the Mother Tongue.* New York: John Wiley, 1996.

Sikora, Martine, et al. "Ancient Genomes Show Social and Reproductive Behavior of Early Upper Paleolithic Foragers." *Science* (November 2017).

Skoglund, Pontus, and David Reich. "A Genomic View of the Peopling of the Americas." *Current Opinion in Genetics & Development* (December 2016): 27–35.

Tamm, Erika, et al. "Beringian Standstill and Spread of Native American Founders." *PLoS ONE* (September 2007).

Who Drank the First Beer?

Bostwick, William. *The Brewer's Tale: A History of the World According to Beer.* New York: W. W. Norton, 2014.

Braidwood, Robert J., et al. "Did Man Once Live by Beer Alone?" *American Anthropologist* 55, no. 4 (October 1953).

Gallone, Brigida, et al. "Domestication and Divergence of *Saccharomyces cerevisiae* Beer Yeasts." *Cell* (September 2016).

Hayden, Brian, et al. "What Was Brewing in the Natufian? An Archaeological Assessment of Brewing Technology in the Epipaleolithic." *Journal of Archaeological Method and Theory* (March 2013).

Hobhouse, Henry. *Seeds of Change: Six Plants That Transformed Mankind.* Los Angeles, CA: Counterpoint, 2005.

Katz, Solomon, and Mary Voigt. "Bread and Beer: The Early Use of Cereals in the Human Diet." *Expeditions* 28, no. 2 (1986): 23–35.

Legras, J. L., et al. "Bread, Beer and Wine: Saccharomyces cerevisiae Diversity Reflects Human History." *Molecular Ecology* (May 2007).

Moore, A. M. T., et al. "The Excavation of Tell Abu Hureyra in Syria: A Preliminary Report." *Proceedings of the Prehistoric Society* (1975).

National Geographic. "What the World Eats." National Geographic website, April 2019. https://www.nationalgeographic.com/what-the-world-eats.

Paulette, Tate, and Michael Fisher. "Potent Potables of the Past: Beer and Brewing in Mesopotamia." *Ancient Near East Today* 5, no. 4 (April 2017).

Rodger, N.A.M. *The Command of the Ocean: A Naval History of Britain 1649–1815.* New York: Penguin Books, 2006.

Smalley, John, and Michael Blake. "Sweet Beginnings: Stalk Sugar and the Domestication of Maize." *Current Anthropology* (December 2003).

Standage, Tom. *A History of the World in 6 Glasses*. New York: Bloomsbury, 2005.

Yaccino, Steven. "For Its Latest Beer, a Craft Brewer Chooses an Unlikely Pairing: Archaeology." *New York Times,* June 17, 2013.

Who Performed the First Surgery?

Albanèse, J., et al. "Decompressive Craniectomy for Severe Traumatic Brain Injury: Evaluation of the Effects at One Year." *Critical Care Medicine* (October 2003).

Alt, Kurt W., et al. "Evidence for Stone Age Cranial Surgery." *Nature* (June 1997).

Butterfield, Fox. "Historical Study of Homicide and Cities Surprises the Experts." *New York Times,* October 23, 1994.

Faria, Miguel A. "Neolithic Trepanation Decoded: A Unifying Hypothesis: Has the Mystery As to Why Primitive Surgeons Performed Cranial Surgery Been Solved?" *Surgical Neurology International* (May 2015).

Henshcen, Folke. *The Human Skull: A Cultural History*. Santa Barbara, CA: Praeger Publishers, 1965.

Hershkovitz, I. "Trephination: The Earliest Case in the Middle East." *Mitekufat Haeven: Journal of the Israel Prehistoric Society* (1987): 128–35.

Lv, Xianli, et al. "Prehistoric Skull Trepanation in China." *World Neurosurgery* (December 2013): 897–99.

Meyer, Christian, et al. "The Massacre Mass Grave of Schöneck-Kilianstädten Reveals New Insights into Collective Violence in Early Neolithic Central Europe." *PNAS* (September 2015): 11217–222.

Olalde, Iñigo, et al."A Common Genetic Origin for Early Farmers from Mediterranean Cardial and Central European LBK Cultures." *Molecular Biology and Evolution* (December 2015): 3132–42.

Petrone, Pierpaolo, et al. "Early Medical Skull Surgery for Treatment of Post-Traumatic Osteomyelitis 5,000 Years Ago." *PloS One* (May 2015).

Prioreschi, Plinio. *A History of Medicine: Primitive and Ancient Medicine*. N.p.: Horatius Press, 2002.

Rudgley, Richard. *The Lost Civilizations of the Stone Age*. New York: Free Press,1999.

Sigerist, Henry. *A History of Medicine Volume 1: Primitive and Archaic Medicine*. Oxford, UK: Oxford University Press, 1951.

Verano, John. *Holes in the Head: The Art and Archaeology of Trepanation in Ancient Peru*. Cambridge, MA: Harvard University Press, 2016.

Watson, Traci. "Amazing Things We've Learned from 800 Ancient Skull Surgeries." *National Geographic,* June 30, 2016.

Who First Rode the Horse?

Anthony, David. *The Horse, the Wheel, and Language: How Bronze-Age Riders from the Eurasian Steppes Shaped the Modern World.* Princeton, NJ: Princeton University Press, 2007.

Anthony, David, and Dorcas Brown. "Horseback Riding and Bronze Age Pastoralism in the Eurasian Steppes." In *Reconfiguring the Silk Road*, edited by Victor Mair and Jane Hickman. Philadelphia, PA: University of Pennsylvania Press, for the Museum of Archaeology and Anthropology, 2014.

Chang, Will, et al. "Ancestry-Constrained Phylogenetic Analysis Supports the Indo-European Steppe Hypothesis." *Language* (January 2015).

Diamond, Jared. *Guns, Germs, and Steel: The Fates of Human Societies.* New York: W. W. Norton, 1999.

Haak, Wolfgang, et al. "Massive Migration from the Steppe Is a Source for Indo-European Languages in Europe." *Nature* (June 2015).

Olsen, Sandra. "Early Horse Domestication on the Eurasian Steppe." *Documenting Domestication: New Genetic and Archaeological Paradigms* (2006): 245–69.

Outram, Alan K., et al. "The Earliest Horse Harnessing and Milking." *Science* (March 2009).

Who Invented the Wheel?

Anthony, David. *The Horse, the Wheel, and Language: How Bronze-Age Riders from the Eurasian Steppes Shaped the Modern World.* Princeton, NJ: Princeton University Press, 2007.

Anthony, David W., and Don Ringe. "The Indo-European Homeland from Linguistic and Archaeological Perspectives." *Annual Review of Linguistics* 1 (2015): 199–219.

Bouckaert, Remco, et al. "Mapping the Origins and Expansion of the Indo-European Language Family." *Science* (2012).

Callaway, Ewen. "Bronze Age Skeletons Were Earliest Plague Victims." *Nature* (October 2015).

Charnay, Désiré. *The Ancient Cities of the New World.* Cambridge, UK: Cambridge University Press, 2013.

Hassett, Janice, et al. "Sex Differences in Rhesus Monkey Toy Preferences Parallel Those of Children." *Hormones and Behavior* (August 2008): 359–64.

Rasmussen, Simon, et al. "Early Divergent Strains of Yersinia pestis in Eurasia 5,000 Years Ago." *Cell* (October 2015).

Reinhold, Sabine, et al. "Contextualising Innovation: Cattle Owners and Wagon Drivers in the North Caucasus and Beyond." In *Appropriating Innovations: Entangled Knowledge in Eurasia, 5000–150 BCE*. Oxford, UK: Oxbow Books, 2017.

Vogel, Steven. *Why the Wheel Is Round: Muscles, Technology, and How We Make Things Move*. Chicago, IL: University of Chicago Press, 2016.

Williams, Christina, and Kristen Pleil. "Toy Story: Why Do Monkey and Human Males Prefer Trucks?" *Hormones and Behavior* (May 2008).

Who Was the Murderer in the First Murder Mystery?

Bowles, Samuel, et al. "Did Warfare Among Ancestral Hunter-Gatherers Affect the Evolution of Human Social Behaviors?" *Science* (June 5, 2009): 1293–98.

Brennan, Bonnie, and David Murdock. *Nova:* "Iceman Reborn." Produced and directed by Bonnie Brennan. Arlington, VA: PBS, February 17, 2016.

Bulger, Burkhard. "Sole Survivor." *New Yorker*, February 14, 2005.

Butterfield, Fox. "Historical Study of Homicide and Cities Surprises the Experts." *New York Times*, October 23, 1994.

Feltman, Rachel. "What Was Otzi The Iceman Wearing When He Died? Pretty Much the Entire Zoo." *Washington Post*, August 18, 2016.

Hanawalt, Barbara A. "Violent Death in Fourteenth and Early Fifteenth-Century England." *Comparative Studies in Society and History* 18, no. 3 (July 1976): 297–320.

Müller, Wolfgang, et al. "Origin and Migration of the Alpine Iceman." *Science* (2003): 862–66.

Nordland, Rod. "Who Killed the Iceman? Clues Emerge in a Very Cold Case." *New York Times*, March 26, 2017.

Oeggl, Klaus, et al. "The Reconstruction of the Last Itinerary of 'Ötzi,'" the Neolithic Iceman, by Pollen Analyses from Sequentially Sampled Gut Extracts." *Quaternary Science Reviews* 26 (2007): 853–61.

Pinker, Steven. *The Better Angels of Our Nature: Why Violence Has Declined*. New York: Viking Press, 2011.

United Nations Office on Drugs and Crime. *Global Study on Homicide 2013* (2014).

Wrangham, Richard W., et al. "Comparative Rates of Violence in Chimpanzees and Humans." *Primates* (January 2006).

Who Was the First Person Whose Name We Know?

Alster, Bendt. *Proverbs of Ancient Sumer: The World's Earliest Proverb Collections.* Potomac, MD: Capital Decisions, University Press of Maryland, 1997.

Bohn, Lauren E. "Q&A: 'Lucy' Discoverer Donald C. Johanson." *Time*, March 4, 2009.

Devlin, Keith. *The Math Gene: How Mathematical Thinking Evolved and Why Numbers Are Like Gossip.* New York: Basic Books, 2000.

Fischer, Steven Roger. *A History of Writing.* London: Reaktion Books, 2001.

Graeber, David. *Debt: The First 5000 Years.* Brooklyn, NY: Melville House, 2014.

Harari, Yuval Noah. *Sapiens: A Brief History of Humankind.* New York: Harper Perennial, 2015.

Haub, Carl. "How Many People Have Ever Lived on Earth?" *Population Today* (February 1995).

Nissen, Hans, Peter Damerow, and Robert Endlund. *Archaic Bookkeeping.* Chicago, IL: University of Chicago Press, 1994.

Renn, Jürgen. "Learning from Kushim About the Origin of Writing and Farming: Kushim—Clay Tablet (c. 3200–3000 BCE), Erlenmeyer Collection," (2014).

Sagona, Tony. "The Wonders of Ancient Mesopotamia: How Did Writing Begin?" The Wonders of Ancient Mesopotamia Lecture Series, University of Melbourne. Presented by Museum Melbourne, 2012.

Schmandt-Besserat, Denise. *How Writing Came About.* Austin: University of Texas Press, 1997.

Scott, James C. *Against the Grain: A Deep History of the Earliest States.* New Haven, CT: Yale University Press, 2017.

Shoumatoff, Alex. *The Mountain of Names: A History of the Human Family.* New York: Kodansha International, 1985.

Stadler, Friedrich. *Integrated History and Philosophy of Science: Problems, Perspectives, and Case Studies.* New York: Springer, 2017.

Who Discovered Soap?

Adams, Robert. *Heartland of Cities: Surveys of Ancient Settlement and Land Use on the Central Floodplain of the Euphrates.* Chicago, IL: University of Chicago Press, 1981.

Bhanoo, Sindya. "Remnants of an Ancient Kitchen Are Found in China." *New York Times*, June 28, 2012.

Curtis, John. "Fulton, Penicillin and Chance." *Yale Medicine* (Fall/Winter 1999/2000).

Curtis, V., and S. Cairncross. "Effect of Washing Hands with Soap on Diarrhoea Risk in the Community: A Systematic Review." *Lancet Infectious Diseases* (May 2003): 275–81.

Dunn, Robb. *Never Home Alone: From Microbes to Millipedes, Camel Crickets, and Honeybees, the Natural History of Where We Live.* New York: Basic Books, 2018.

Konkol, Kristine, and Seth Rasmussen. "An Ancient Cleanser: Soap Production and Use in Antiquity." *Chemical Technology in Antiquity* (November 2015): 245–66.

Luby, Stephen, et al. "The Effect of Handwashing at Recommended Times with Water Alone and with Soap on Child Diarrhea in Rural Bangladesh: An Observational Study." *PLoS Medicine* (June 2011).

———. "Effect of Handwashing on Child Health: A Randomized Controlled Trial." *The Lancet* (July 2005): 225–33.

Levey, Martin. "Dyes and Dyeing in Ancient Mesopotamia." *Journal of Chemical Education* (December 1955).

———. "The Early History of Detergent Substances: A Chapter in Babylonian Chemistry." *Journal of Chemical Education* (October 1954).

Nemet-Nejat, Karen Rhea. *Daily Life in Ancient Mesopotamia.* Westport, CT: Greenwood Press, 1998.

———. "Women's Roles in Ancient Mesopotamia." In *Women's Roles in Ancient Civilizations: A Reference Guide,* edited by Bella Vivante. Westport, CT: Greenwood Press. 1999.

Sallaberger, Walther. "The Value of Wool in Early Bronze Age Mesopotamia. On the Control of Sheep and the Handling of Wool in the Presargonic to the Ur III Periods (c. 2400 to 2000 BC)." In *Wool Economy in the Ancient Near East and the Aegean: From the Beginnings of Sheep Husbandry to Institutional Textile Industry,* edited by Catherine Breniquet and, Cécile Michel (Hg). Oxford, UK: Oxbow Books, 2014.

Saxon, Wolfgang. "Anne Miller, 90, First Patient Who Was Saved by Penicillin." *New York Times,* June 9, 1999.

Tager, Morris. "John F. Fulton, Coccidioidomycosis, and Penicillin." *Yale Journal of Biology and Medicine* (September 1976): 391–98.

Wright, Rita. "Sumerian and Akkadian Industries: Crafting Textiles." In *The Sumerian World,* edited by H. E. W. Crawford. New York: Routledge Press, 2013.

Who Caught the First Case of Smallpox?

Abokor, Axmed Cali. *The Camel in Somali Oral Traditions.* Mogadishu, Somalia: Somali Academy of Sciences and Arts, 1987.

Babkin, Igor, and Irina Babkina. "A Retrospective Study of the Orthopoxvirus Molecular Evolution." *Infection, Genetics and Evolution* (2012): 1597–1604.

Barquet, Nicolau. "Smallpox: The Triumph over the Most Terrible of the Ministers of Death." *Annals of Internal Medicine* 128 (1997).

Broad, William J., and Judith Miller. "Report Provides New Details of Soviet Smallpox Accident." *New York Times,* June 15, 2002.

Bulliet, Richard W. *The Camel and the Wheel.* New York: Columbia University Press, 1990.

Esposito, J. J. "Genome Sequence Diversity and Clues to the Evolution of Variola (Smallpox) Virus." *Science* (2006): 807–12.

Foege, William H. *House on Fire: The Fight to Eradicate Smallpox.* Berkeley and Los Angeles: University of California Press, 2011.

Goldewijk, Klein, et al. "Long-Term Dynamic Modeling of Global Population and Built-Up Area in a Spatially Explicit Way." *Holocene* (2010): 565–73.

Gubser, Caroline, and Geoffrey Smith. "The Sequence of Camelpox Virus Shows It Is Most Closely Related to Variola Virus, the Cause of Smallpox." *Journal of General Virology* (2002): 855–72.

Henderson, D. A. *Smallpox: The Death of a Disease.* Amherst, NY: Prometheus Books, 2009.

Needham, Joseph. *China and the Origins of Immunology.* Hong Kong: Centre of Asian Studies, University of Hong Kong, 1980.

Prankhurst, Richard. *The Ethiopian Borderlands: Essays in Regional History from Ancient Times to the End of the 18th Century.* Trenton, NJ: Red Sea Press, 1997.

Wilson, Bee. *Consider the Fork: A History of How We Cook and Eat.* New York: Basic Books, 2012.

Wolfe, Nathan. *The Viral Storm: The Dawn of a New Pandemic Age.* New York: Times Books, 2011.

Who Told the First Joke We Know?

Alster, Bendt. *Proverbs of Ancient Sumer: The World's Earliest Proverb Collections.* Potomac, MD: Capital Decisions, University Press of Maryland, 1997.

Beard, Mary. *Laughter in Ancient Rome: On Joking, Tickling, and Cracking Up.* Berkeley and Los Angeles: University of California Press, 2014.

Foster, Benjamin. "Humor and Cuneiform Literature." *Journal of Ancient Near Eastern Literature* (1974).

Friend, Tad. "What's So Funny?" *New Yorker,* November 11, 2002.

George, Andrew, trans. *The Epic of Gilgamesh.* New York: Penguin Classics, 2003.

Hurley, Matthew, Daniel Dennett, and Reginald Adams Jr. *Inside Jokes.* Boston, MA: MIT Press, 2011.

Matuszak, Jana. "Assessing Misogyny in Sumerian Disputations and Diatribes." *Gender and Methodology in the Ancient Near East* (2018): 259–72.

Mohr, Melissa. *Holy Shit: A Brief History of Swearing.* Oxford, UK: Oxford University Press, 2013.

Weems, Scott. *Ha!: The Science of When We Laugh and Why.* New York: Basic Books, 2014.

Who Discovered Hawaii?

Bae, Christopher J., et al. "On the Origin of Modern Humans: Asian Perspectives." *Science* (December 8, 2017).

Buckley, Hallie R. "Scurvy in a Tropical Paradise? Evaluating the Possibility of Infant and Adult Vitamin C Deficiency in the Lapita Skeletal Sample of Teouma, Vanuatu, Pacific Islands." *International Journal of Paleopathology* (2014).

Callaghan, Richard, and Scott M. Fitzpatrick. "Examining Prehistoric Migration Patterns in the Palauan Archipelago: A Computer Simulated Analysis of Drift Voyaging." *Asian Perspectives* 47, no. 1 (2008).

———. "On the Relative Isolation of a Micronesian Archipelago During the Historic Period: The Palau Case-Study." *International Journal of Nautical Archaeology* (2007): 353–64.

Collerson, Kenneth, and Marshall Weisler. "Stone Adze Compositions and the Extent of Ancient Polynesian Voyaging and Trade." *Science* (2007).

Fischer, Steven Rodger. *A History of the Pacific Islands.* London: Palgrave Macmillan, 2002.

Fitzpatrick, Scott M., and Richard Callaghan. "Examining Dispersal Mechanisms for the Translocation of Chicken *(Gallus gallus)* from Polynesia to South America." *Journal of Archaeological Science* 36 (2009): 214–23.

———. "Magellan's Crossing of the Pacific." *Journal of Pacific History* 43, no. 2 (2008): 145–65.

Hershkovitz, Israel, et al. "Levantine Cranium from Manot Cave (Israel) Foreshadows the First European Modern Humans." *Nature* (2015).

Howe, K. R., ed. *Vaka Moana, Voyages of the Ancestors: The Discovery and Settlement of the Pacific.* Honolulu: University of Hawaii Press, 2007.

Kirch, Patrick Vinton. *A Shark Going Inland Is My Chief: The Island Civilization of Ancient Hawai'i.* Berkeley and Los Angeles: University of California Press, 2012.

Montenegro, Alvaro, et al. "From West to East: Environmental Influences on the Rate and Pathways of Polynesian Colonization." *Halocene* 24, no. 2 (2014): 242–56.

National Geographic Learning. "Beyond the Blue Horizon." In *Archaeology: National Graphic Learning Reader Series.* Boston, MA: Cengage Learning, 2012.

O'Connell, J. F., et al. "Pleistocene Sahul and the Origins of Seafaring." In *The Global Origins and Development of Seafaring,* edited by Katie Boyle and Atholl Anderson. Cambridge, UK: McDonald Institute for Archaeological Research, 2010.

Reich, David. *Who We Are and How We Got Here: Ancient DNA and the New Science of the Human Past.* New York: Pantheon Books, 2018.

Sobel, Dava. *Longitude: The True Story of a Lone Genius Who Solved the Greatest Scientific Problem of His Time.* New York: Bloomsbury, 2007.

Wayfinders: A Pacific Game. Produced and directed by Gail Evenari. Arlington, VA: PBS, 1999.